Gotha
P.60

David Myhra

Schiffer Military History
Atglen, PA

Book Design by Ian Robertson.

Copyright © 2001 by David Myhra.
Library of Congress Control Number: 2001089995

Printed in China.
ISBN: 0-7643-1399-1

We are interested in hearing from authors with book ideas on related topics.

Published by Schiffer Publishing Ltd.
4880 Lower Valley Road
Atglen, PA 19310
Phone: (610) 593-1777
FAX: (610) 593-2002
E-mail: Schifferbk@aol.com.
Visit our web site at: www.schifferbooks.com
Please write for a free catalog.
This book may be purchased from the publisher.
Please include $3.95 postage.
Try your bookstore first.

In Europe, Schiffer books are distributed by:
Bushwood Books
6 Marksbury Avenue
Kew Gardens
Surrey TW9 4JF
England
Phone: 44 (0) 20 8392-8585
FAX: 44 (0) 20 8392-9876
E-mail: Bushwd@aol.com.
Free postage in the UK. Europe: air mail at cost.
Try your bookstore first.

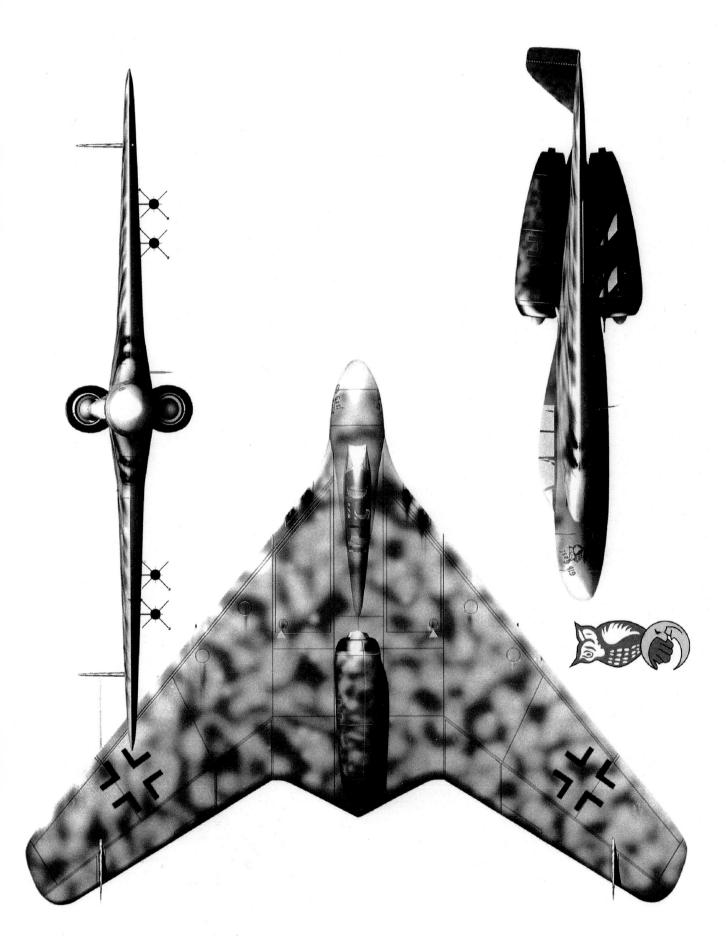

Gothaer Go Projekt 60B

This is a true story of an attempted murder...the killing off of a competitor's flying machine that the *Reichsluftfahrtministerium (RLM, or German Air Ministry)* had approved for series production (*Horten Ho 229*) which they had been ordered to construct under license (*Gothaer Waggonfabrik AG*), and replacing it with one of their own proposed designs (*Gothaer Go P.60B*). No one at *Gothaer* paid any price for this attempted murder, although there is a "smoking gun" (*Gothaer Go P.60B*). In addition, there is considerable circumstantial evidence. The author would like the reader to know at this time that his and other's opinions expressed about this attempted mugging are to be considered opinions and not matters of fact, although the facts are considerable, as will be seen. This said, let's get on with the fascinating story of the *Gothaer Go P.60B* verses the *Horten Ho 229*.

Gothaer Waggonfabrik AG-Gotha, or "*GWK*" (the words "*Gothaer*" and "*Gotha*" were frequently used to identify the *Gothaer Waggonfabrik AG* and both are correct), had been instructed by the *Luftwaffe Quartermaster General*-Berlin of the German Aviation

Ministry, in August 1944 that it was to take the prototype, twin *Jumo 004B* powered all-wing flying machine *Horten Ho 9 V2* and manufacture it in series production. For series production, the *RLM* said that the *Horten Ho 9 V2* would be known as the *Horten Ho 229*. *Dr.-Ing. Rudolf Göthert* [1912-1973] was the chief aerodynamicist for *Gothaer Waggonfabrik AG*, having joined the firm in early 1942 after leaving *LFA's Institut für Aerodynamik (Luftfahrt Forschungs Anstalt-* Braunschweig/Völkenrode), or Aircraft Research Establishment-Institute for Aerodynamics. Prior to joining *LFA* in 1937, *Rudolf Göthert* had been employed by *DVL (Deutsche Versuchsanstalt für Luftfahrt-*Berlin/Adlershof), or German Research Center for Aviation, since 1935. His doctoral dissertation at the *Hannover Technische Hochschule* was titled "Systematic Investigation of Wings with Flaps and Tabs." It basically covered his research work at the *LFA*-Braunschweig/Völkenrode between 1935 and 1940, and he received his *Ph.D.* (mathematics and physics) in 1940. *Göthert* continued working at *LFA* until 1942 when he took the position of chief aerodynamicist at *Gothaer Waggonfabrik AG*-Gotha.

Gothaer Waggonfabrik's **manufacturing facilities at the city of Gotha as seen in the mid 1930s.**

Gothaer Waggonfabrik's **modern looking but functional corporate headquarters building in Berlin. Seen in the mid 1930s.**

Gothaer Waggonfabrik AG received a contract to construct up to twenty *Horten Ho 229s* from *Artur Eschenhauer* at the *RLM's Luftwaffe* Quartermasters Office about August 1944. After *Rudolf Göthert* had studied the design plans for what the *RLM* had designated as the *Horten Ho 229*, *Dr. Rudolf Göthert* began, in effect, telling the *RLM* that the *Horten Ho 229* was no damn good. If the *Lutfwaffe* really wanted an all-wing heavy pursuit machine (day fighter and a night fighter) piloted by a crew of one or two, and it appeared that the *RLM* was very interested, then he and his friends from *LFA* and *DVL* could, through scientific wind tunnel-tested design, give them a more advanced all-wing superior to the *Horten Ho 229* they were currently making ready for serial production. *Dr. Rudolf Göthert* was calling his *Horten Ho 229* killer/replacement the *Gothaer Go Projekt 60B*. His twin brother, *Dr.-Ing. Berhard Göthert,* had already been conducting aeronautical wind tunnel research on the *Horten Ho 9 V1* sailplane in 1944 at the prestigious *DVL* where he was employed in wind tunnel research.

Berhard Göthert told this author in a 1981 telephone interview, that he did not think that it was his brother's intention to kill the *Horten* brother's turbojet powered all-wing *Horten Ho 229*, but that he may have gotten caught up in the spirt of the time, or "Zeitgeist." *Berhard Göthert* described how management at *Gothaer*

Waggonfabrik AG had come to the conclusion that future aircraft design belonged to the all-wing plan form. How *Gothaer* came to this conclusion he personally did not know. *Rudolf Göthert* had told his American Army Air Force Intelligence (*AAAFI*) interrogators, under *General George C. McDonald*, post war that management at *Gothaer Waggonfabrik* had read about the American *Jack Northrop's* all-wing flying machine in a 1942 issue of *Inter-Avia* magazine. A photo of *Northrop's* all-wing flying machine was included in the article. *Gothaer's* management and the *RLM*, as well as the *Horten* brothers and other aviation specialists throughout Germany believed that *Jack Northrop* had somehow perfected the all-wing regarding its lateral instability problems. Germans were believing that America would soon put all-wing bombers and fighters into the air over Germany, and they had to be ready with their own all-wing flying machines to counter the expected American threat. *Gothaer Waggonfabrik* wanted to become the German leader in the field of all-wing research, design, and construction. When they went looking for a person to spear-head their all-wing program...they found and later hired *Dr.-Ing. Rudolf Göthert* away from *LFA*.

Although original aircraft design was not *Gothaer's* main field of aviation activity, they wanted to become involved in the design and construction of all forms of all-wing aircraft...day fighters, night

Rudolf Göthert's first proposed twin turbojet powered high-speed fighter design was his all-wing Gothaer Go P.60A. The initial design was an all-wing with two upper and under fixed stabilizing fins at the wing tips. Were it not for the over and under mounting of the BMW 003A turbojet engines, the proposed machine looks a lot like the Horten Ho 229...especially its center section. Göthert had given up on the fixed stabilizing fins, and this machine would have been fitted with wing tip vertical stabilizers with a hinged rudder. No longer than an all-wing...the prone piloted Gothaer Go P.60A would have evolved into a tailless flying machine. Scale model and photographed by Reinhard Roser.

Left: Dr.-Ing. Rudolf Göthert, chief aviation aerodynamicist at Gothaer. He obtained all his wind tunnel experience at LFA, while his brother...Dr.-Ing. Berhard Göthert, was a wind tunnel expert at DVL. Both Göthert brothers had influence and respect within the Third Reich's aviation research laboratories. Rudolf Göthert left LFA and joined the highly conservative Gothaer Waggonfabrik in 1942. Perhaps, like his colleague Hans Multhopp, he too wanted to see his theories of tailless aviation design transformed into metal, and Gothaer Waggonfabrik wanted to get into the research, design, and construction of tailless and all-wing flying machines, too.

Dr. Rudolf Göthert's all-wing *Gothaer Go P.60A* as seen from its port side. The cockpit cabin canopy would have been flush with the surrounding center section. Unique among forthcoming aviation fighter designs submitted to the *RLM* in 1944/1945, was *Göthert's* design plans for placing the machine's two man crew in prone flying positions. No *Third Reich* fighter had had a prone flying and fighting positions. What about pilot efficiency and visibility? An untried concept for a fighter from *Rudolf Göthert* at *Gothaer Waggonfabrik*. Scale model and photographed by *Reinhard Roeser*.

fighters, heavy bombers, and ultra long range bombers. *Berhard Göthert* described to this author how *Gothaer Waggonfabrik* had initially talked with the *Horten* brothers about getting into all-wing research and design. (This author is not aware of any such talks between the *Horten* brothers and *Gothaer Waggonfabrik*.) Nevertheless, about 1942 *Gothaer Waggonfabrik* began looking around for a capable all-wing research and design person. They found and hired *Dr.-Ing. Rudolf Göthert* away from *LFA*-Braunschweig/ Völkenrode. It appears that it may have been *Gothaer's* intention to obtain the series production contract for the *Horten Ho 229* merely for the experience, while having *Dr. Rudolf Göthert* apply the all-wing concept to other forms of aircraft.

Dr. Berhard Göthert told this author that he believed his twin brother, after he saw and studied the engineering drawings for the *Horten Ho 229*, which *Gothaer* had been ordered to construct in series, immediately felt that this design was out of date and that he, his wind tunnel colleagues at *DVL* and *LFA*, and *Gothaer Waggonfabrik* could give the *RLM* a better design right then to replace the poorly designed *Horten Ho 229*.

"My brother believed that the *Horten Ho 229* he was looking at was old fashioned, old thinking, and based on wind tunnel tests

This fighter interceptor seen from its rear port side is what the *Gothaer Go P.60A* evolved into...the tailless *Gothaer Go P.60B*. Dr. *Göthert's* 2nd version featured the pilot and copilot seated in the traditional/conventional manner, that is, upright and in tandem. But *Rudolf Göthert* also made huge engineering changes. *Rudolf Göthert* got rid of the fixed stabilizing fins on his earlier version and definitely went to a vertical stabilizer on each wing tip with an attached hinged rudder. He said that the stabilizer and rudder were required features on an all-wing idea if one were to get the needed directional stability which a fighter/ interceptor required. Of course, the *Gothaer Go P.60B* was no longer an all-wing but a tailless flying machine. Scale model and photographed by *Reinhard Roeser*.

A lot had changed on the *Gothaer Go P.60B* from the *Go P.60A*. Aside from being equipped with twin *Heinkel-Hirth HeS 011* turbojet engines, the design fad/trend of twin vertical stabilizers with hinged rudders had become the signature design of Dr. *Rudolf Göthert*. Elsewhere it could be found on other advanced designs in the *Third Reich* 1944/1945. For example, the proposed *Junkers Ju EF 128* design featured twin vertical stabilizers with hinged rudders. Scale model and photographed by *Reinhard Roeser*.

The proposed E*F 128* by *Junkers* with its twin oversized vertical stabilizers each with a hinged rudder. Courtesy of *Dan Johnson* and *Daniele Sabatini, Luftwaffe Project Aircraft #1, Junkers EF 128*, Self published, 1999.

both of us had done at *DVL* and *LFA*. We both sincerely believed that the *Horten Ho 229* could be made more efficient, and with a few design changes could thus have a higher level of performance as a war machine. That higher level, more modern fighting machine was to be the *Gothaer Go P.60*."

It appears that *Rudolf Göthert,* based on what his twin brother *Berhard* told this author in 1981, believed that the *Luftwaffe* in 1944/1945 needed the most efficient and capable fighting machine which could be designed right away. So strongly did *Rudolf Göthert* believe this that he felt that the *Horten Ho 229*, based on old concepts and wrong experience coming from the *Horten's Rhön/Wasserkuppe* sailplane days, that the *Horten* brothers had to give way to a more scientific analysis of fighter airplane design obtained through wind tunnel results. *Göthert* believed that the *Horten* brothers' experience in all-wing sailplanes did not apply to a heavy turbojet powered all-wing fighter. In order to put the all-wing fighting machine on a performance level with the *Messerschmitt Me 262*, the *Horten Ho 229* had to leave the stage and embrace *Rudolf Göthert's* tailless *Gothaer Go P.60*. Dr. *Berhard Göthert* suggested that the reason why *Gothaer's* assembly work on the *Horten Ho 229 V3* through *V6* had progressed so slowly was because *Gothaer Waggonfabrik* was purposely dragging their feet and hoping that the *RLM* would cancel their contract with the *Horten* brothers in favor of *Gothaer's Go P.60B*.

The academic-oriented researchers at *DVL* and *LFA* had power, and they had connections throughout the entire German aviation community. All the important aviation constructors in Germany respected the work of the *LFA* and *DVL*, followed their work, and applied their findings to new aircraft designs. They included *Willy Messerschmitt* and *Woldemar Voigt* [*Messerschmitt AG*], *Kurt Tank, Hans Multhopp* and *Gotthold Mathias* [*Focke-Wulf*], *Richard Vogt* [*Blohm & Voss*], *Walter Georgii* and *Felix Kracht* [*DFS*], and others from the German aviation industry. But these aircraft designers did not include the *Horten* brothers, *Walter* and *Reimar*. The *Horten* brothers had no formal education in aeronautics. *Walter* [1913-1998] had not even gone on to college after graduating high school. *Reimar* [1915-1994] had gone to college, off and on, studying mathematics. He earned his *PhD* post war. All of the *Horten* brother's aircraft design experience was self-taught, coming about through their sailplanes...trial and error...just like *Alexander Lippisch, Willy Messerschmitt,* and *Ernst Heinkel* in their early years. Yet, despite the *Horten's* academic shortcomings, they still had managed to design, construct, and fly a twin turbojet-powered all-wing aircraft. This feat put the *Horten* brothers in very exclusive company, which only included those people who designed the *Messerschmitt Me 262, Arado Ar 234B/Ar 234C,* and the *Heinkel He 280* and *He 162*. So, not only did the *Horten* brothers manage to get their twin turbo-

An overhead port side view of *Göthert's Gothaer Go P.60B*, the tailless machine which *Rudolf Göthert*, his brother *Berhard Göthert*, and their other aviation research center colleagues would use to kill the *Horten Ho 229* by those upstart, self educated, name dropping, and boot-licking *Horten* brothers...*Walter* and *Reimar*. Scale model and photographed by *Reinhard Roeser*.

This is what the fuss was all about...the H*orten Ho 229* fighter/interceptor seen here from its port side with its twin *Jumo 004B* engines under full thrust. *Rudolf Göthert* and his colleagues claimed that the *Horten Ho 229* would never achieve satisfactory directional stability without some form of a vertical stabilizer with an attached rudder. Then again, they claimed the *Horten* machine was too heavy and its center-of-gravity was out of wack...especially during its landing approach. *Rudolf Göthert* privately called the *Horten* all-wing a "pilot killer," too. Logical, rational, and practical thinking meant that its serial production had to be stopped. Digital image by *Mario Merino*.

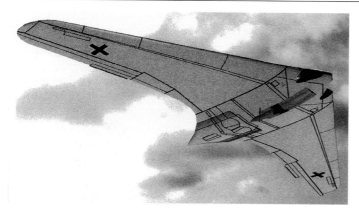

The *Göthert* twins and their aviation wind tunnel experts/friends claimed that they were opposing the *Horten Ho 229* because it was the wrong machine for turning around the *Third Reich's* failing war effort. In addition, poorly trained young pilots would be unable to effectively compete in combat with this machine. *Göthert* claimed that their opposition to the *Horten Ho 229* was not due to any jealousy they may have had of the *Horten* brothers' successes with the *RLM*. Digital image by *Mario Merino*.

A tandem-seat *Horten Ho 229* on the tarmac as seen from above, behind, and featuring its port side. It probably would not have carried a *swastika* due to a lack of a vertical surface. Its upper wing surface would probably have been camouflaged with either *81 Brown-Violet, 82 Bright Green,* or *83 Dark Green*. Digital image by *Mario Merino*.

jet *Horten Ho 9 V2* airborne, but they were also able to convince *Reichsmarshall Hermann Göring*, leader of the *Luftwaffe*, to put their prototype *Horten Ho 9 V2* into series production as the *Horten Ho 229*. How they managed all this, *DVL* and its supporters did not quite understand. But they knew that the *Horten* brothers were not members of *DVL or LFA's* community, nor had their *Horten Ho 9 V2* ever been wind tunnel tested. So *DVL* and *LFA*, and pretty much everyone else connected to these prestigious aviation research centers were jealous over the success of the *Autobahn* Vehicle Maintenance workshop-built all-wing *Horten Ho 9 V2*. The fact that it had been selected for serial production as a day fighter designated as the *Horten Ho 229* from the rag-tag operation of the *Horten* brothers was absurd. German aviation research experts told their contacts at the *RLM* that colleague *Dr. Rudolf Göthert* could do better...much better given the *Third Reich's* desperate need for a

superior fighter/interceptor aircraft by 1944 to counter the deadly daily threat of those *Boeing B-17s* especially, flying in their defensive "box" formations. And then there were those free-roaming *North American P.51 "Mustangs"* searching the German countryside to hit anything that moved on the ground. The best bet to counter these deadly American flying machines, *Göthert* believed, was to select for serial production one of those totally scientific wind-tunnel designed all-wing fighters being created in the wind tunnels of *DVL* and *LFA*, and now being proposed by colleagues and himself at *Gothaer Waggonfabrik AG*. But the *Horten* brothers had their supporters, too; powerful decision-makers far outnumbering any of those pledged to *DVL* and/or *LFA*. The most powerful decision maker in the *Horten's* corner was *Hermann Göring*. In addition, *Walter Horten* had developed a genuine friendship with the brilliant and secretive *Oberst Siegfried Knemeyer*, the powerful leader

A *Horten Ho 229* tandem-seat representing the type *Gothaer Waggonfabrik* was supposed to construct in series as seen on the tarmac from its starboard side. Notice that its wing tip drag rudders have been extended prior to takeoff powered by its twin *Jumo 004B-2* turbojet engines. Digital image by *Mario Merino*.

Two tandem-seat *Horten Ho 229s* on maneuvers. Digital image by *Mario Merino*.

A port side ground level view of a *Horten Ho 229 V7* tandem-seat night fighter equipped with a *FuG 218 "Neptun"* nose mounted radar. Digital image by *Mario Merino*.

of the "*club*" within the *RLM*. *Walter Horten* had been a long time friend of *Oberst Artur Eschenauer*, a senior officer within the *Luftwaffe's Quartermaster General*-Berlin. *Walter Horten* was a genius when it came to winning friends and influencing people. He had systematically built up a wide, strong base of support within the *Luftwaffe*. *Reimar Horten* had the gift of total recall. Even though he did not have a *PhD* in aviation engineering, the shy, withdrawn, secretive, slightly paranoid, and a true believer in the all-wing planform was indeed a genius...perhaps knowing more about all-wing aerodynamics than the whole of *DVL, LFA*, and the *Göthert* brothers combined. It would be a classic power struggle...the back street, self-educated, former rag-tag sailplane builders/fliers from the *Rhön/Wasserkuppe* and senior officers in the *RLM* and *PhD* academics from *DVL* and *LFA* armed with their wind tunnel test data and scientific theories.

What sort of aircraft had *Gothaer Waggonfabrik AG* produced during World War II...mainly pilot training flying machines and

transport gliders. However, during World War I, *Gothaer* had been a producer of multi-piston motored heavy bombers, such as the feared *Gothaer "G,"* and known also as the "*Engländer Fliers*." Post World War I, *Gothaer* was the only German aircraft maker to have all their facilities dismantled and destroyed under the terms of the *Treaty of Versailles*. *Gothaer Waggonfabrik* re-entered the German aircraft industry in the mid-1930s. At first they devoted themselves to the construction of light commercial and training aircraft. They later turned to producing their own glider and powered designs for the *Luftwaffe*. Original *Gothaer*-designed aircraft included:

• *Go 145* - a single engine (*Argus As 10C*), two seat training bi plane;
• *Go 146* - a twin engine (*Hirth HM 508*) 5 or 6 seat light transport;
• *Go 147* - an experimental pterodactyl design for a reconnaissance aircraft designed by *August Kupper* (1905-1937);
• *Go 148* - unused;

A tandem-seat *Horten Ho 229 V7* prototype equipped with a *FuG 218 "Neptun"* radar night fighter version to be constructed in series by the *Gothaer Waggonfabrik*. Digital image by *Mario Merino*.

An overhead view of a *Horten Ho 229 V7*, tandem-seat, *FuG 218 "Neptun"* radar equipped night fighter prototype. Digital image by *Mario Merino*.

A tandem-seat H*orten Ho 229* night fighter equipped with the state-of-the-art *FuG 240* 9 centimetric parabolic scanning internal nose mounted parabolic radar. This flying machine was to also have been built in series for the *Horten* brothers by *Gothaer Waggonfabrik*. Digital image by *Mario Merino*.

A H*orten Ho 229* tandem-seat night fighter equipped with a FuG *240* 9 centrimetric parabolic scanning internal mounted radar as seen about dawn. Digital image by *Mario Merino*.

• *Go 149 -* a single engine (*Argus As 10* or *Argus As 410)*, two seat light training low-wing cabinmonoplane;

• *Go 150 -* a twin engine (*Zundapp Z 9-92*), two seat light communications monoplane;

• *Go 242 -* a medium transport glider high-wing monoplane;

• *Go 244 -* a twin engine (*Gnome-Rhone 14*) transport developed from the *Gothaer Go 242;*

• *Go 345 -* a high wing monoplane transport and training glider;

Gothaer Waggonfabrik AG also produced the following flying machines under contract for the *RLM*:

• *Messerschmitt Me 110 -* a twin motor heavy bomber/destroyer aircraft built by *Gothaer* at its home factory at Gotha beginning in

Above: Two H*orten Ho 229 V7* night fighter prototypes on maneuvers aided by their internal *FuG 240* 9 centimetric parabolic scanning radar. Digital image by *Mario Merino*.

Left: Five single-seat H*orten Ho 229* day fighters of the type which would have been assigned to *Wolfgang Späte*, leader of *JG400*. Digital image by *Mario Merino*.

Gothaer Waggonfabrik AG had been an aviation leader in the "Great War," World War I. Their huge Gotha "G" bombers flew over England, terrorizing its civilian population. Although the amount of bombs they dropped was small, the public was scared of these bi-wing bombers. Gothaer Waggonfabrik was the only Imperial German aviation company mentioned in the Treaty of Versailles. Gothaer's aviation works was totally dismantled and destroyed post World War I...as outlined in the Treaty.

The port side nose/cockpit of the Hans Jacobs design, and which Gothaer Waggonfabrik constructed in series, known as the DF S 230A/ D - 9 passenger transport glider.

1940, with production continuing until the end of the war on May 8th 1945...approximately 6,000 were produced by *Messerschmitt AG* and subcontractors;

• *Focke-Wulf Fw 58* - a twin motor trainer/transport aircraft built by *Gothaer* during the war, production terminated September 1943 after 4,500 were produced...some with floats;

• *Deutsche Forschunganstalt für Segelflug DFS 230* - transport glider built by *Gothaer* during the war, production terminated in June 1941;

Gothaer Waggonfabrik's aviation manufacturing was re-established with the rise to power of Adolf Hitler in 1933. But something had happened to the company's spirit post World War I. They never again became leaders and innovators as they were in their glory days of the "Great War." Oh, they designed and built bi-wing pilot training aircraft, twin engine light transports, and transport gliders designed by Hans Jacobs, formerly of DFS, but they offered their man power and facilities to the RLM as licensed builders of other's aircraft designs. Featured is a poor quality photo of the Gothaer Go 145 C-1. This particular machine, with the radio call code D-IIWS, is carrying a prototype Argus As 014 pulse jet engine during its early flight testing by the Argus engine works.

• *Horten Ho 229 V3* - two machines nearly completed, with several center section frames welded up at war's end (May 8th 1945) out of a contract for 20 machines;

In addition, *Gothaer Waggonfabrik AG* had numerous proposed aircraft projects at war's end. These projects were proposed between 1933 and 1945, and they included:

• *Go P-3.001* - a destroyer aircraft with a double hull, wing span of 525 feet (16 meters), length of 36.4 feet (11.1 meters), hull length of 20 feet (6.1 meters), a height of 6.6 feet (2.0 meters), design year 1935, and powered by *Jumo* or *Damiler-Benz* motors;

• *Go P-3.002* - a larger version of the *Go P-3.001* but featuring a "kinked" wing. Wing span of 55.8 feet (17.0 meters), length of 42 feet (12.8 meters), a hull length of 22.3 feet (6.8 meters), and a height of 9.5 feet (2.9 meters);

• *Go P-4.003* - an airplane for pilot training schools, a further development of the *Go-145* with a stronger motor, 6-cylinder *Argus As* row motor *17A* (already tested in the *Go-145*), wing span of 36 feet (11 meters), length of 31.2 feet (9.5 meters), and a height of 11.5 feet (3.5 meters);

• *Go P-8.001* - a light destroyer with two *Argus As-10C* motors. Wing span of 36.11 feet (11 meters), length of 28.5 feet (8.7 meters), and a height of 8.9 feet (2.7 meters);

• *Go P-9.001* - a training school and sport airplane with two seats and a sequential 4-cylinder *Hirth HM-60R* motor. Wing span of 28.2 feet (8.6 meters), length of 26.2 feet (8.0 meters), and a height of 6.9 feet (2.1 meters);

• *Go P-9.007* - a sport airplane from the year 1939 and powered by a *BMW "X"* motor. Wing span of 32.8 feet (10.4 meters), length of 23.1 feet (7.05 meters), and a height of 6.2 feet (1.9 meters);

Several *Gothaer* built *DFS 230A/D* - 9 passenger transport gliders in the foreground and more beyond.

• *Go P-10.003* - a two seat travel/trip airplane, similar to the *Go-150* but with stronger engines. Wing span of 36.7 feet (11.2 meters), length of 243 feet (7.4 meters), and a height of 6.2 feet (1.9 meters);

• *Go P-11.001* - a two seat training airplane with seats next to one another, 100 horsepower *Hirth HM-504* motor. Wing span of 34.1 feet (10.4 meters), length of 21.2 feet (6.46 meters), height of 5.77 feet (1.76 meters), and generally developed for *NSFK*-competition;

• *Go P-12.001* - a personal travel airplane with twin 100 horse-power engines and similar to the *Go- 241*. Wing span of 39.4 feet (12.0 meters) and a length of 21 feet (6.4 meters);

• *Go P-14.002* - a light destroyer aircraft similar to the *Go P-8.001* and entered by *Gothaer* in competition with the *Messerschmitt Bf 210*. Powered by twin *Argus As-10/410* engines. Wing span of 40.6 feet (12.36 meters), length of 29.4 feet (8.96 meters), and a height of 9.2 feet (2.8 meters);

• *Go P-14.012* - a float plane design entered in competition with the *Arado Ar 196* and similar to the *Go P-14.001*. Wing span of 41 feet (12.5 meters), length of 31.6 feet (9.64 meters), height of 10.2 feet (3.1 meters), and a float length of 11.5 feet (3.5 meters);

• *Go P-16.001* - a light fighter airplane with twin *MK* cannons and similar to the *Go-149* and powered by twin *Argus As 410* motors. Wing span of 27.9 feet (8.5 meters), length of 26.1 feet (7.96 meters), and a height of 7.2 feet (2.2 meters). Forward speed of about 218 miles per hour (350 kilometers/hour);

• *Go P-17.002* - a single seat airplane for flying clubs and powered by a single *Zündapp* motor. Wing span of 24.6 feet (7.5 meters), length of 20.2 feet (6.15 meters), and a height of 5.4 feet (1.65 meters);

• *Go P-20* - a destroyer airplane with three *MG* cannons. Powered by twin *Argus As-10C* motors. Designed in 1938;

• *Go P-21.005* - the further development of the project *Go P-14.002* though with prettier lines, larger wing, and also a stronger motor. Wing span of 50.2 feet (15.3 meters), length of 32.2 feet (9.8 meters), and a height of 6.9 feet (2.1 meters);

• *Go P-35* - a proposed twin piston engine transport with a wing span of 106.5 feet, length of 67.2 feet, and a freight capacity 1,330 cubic feet;

A poor quality photograph of *Gothaer Waggonfabrik's* chief of design...*Albert Kalkert*. This man was responsible for most of the very few new aircraft designs put forth by *Gothaer* between 1933 and 1945. *Kalkert* was responsible for their *Go 145A/B, Go 145D, Go 146, Go 149 Go 150, Go 241, Go 242A/B, Go 345, Kalhert Ka 430,* and along with *Dipl.-Ing Hünerjäger,* the *Gothaer Go 244B/C.*

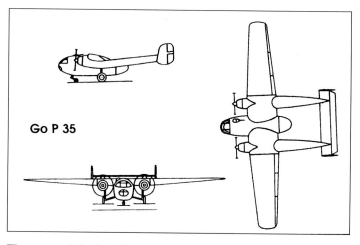

Go P 35

The proposed *Gothaer Go P-35.01* transport, which was to have evolved from the *Gothaer Go 244* assault glider. However, the *Go P-35.01* would have been powered by twin *Bramo 323* piston engines and have had a fixed undercarriage. Protective armament would have included *2xMG 15* cannons...one firing forward and the other firing aft.

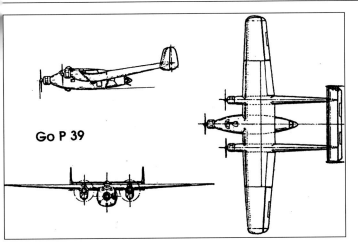

Go P 39

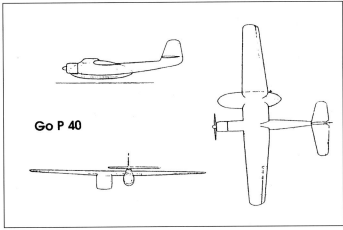

Go P 40

The proposed *Gothaer Go P-39* transport. No specifications are known to exist...only a 3-view design drawing.

The proposed *Gothaer Go P-40B* transport. This flying machine would have had an asymmetrical high-wing plan form and would be powered by a single radial engine. A novel design item was its freight container. Carried beneath its starboard wing, it was designed to be detachable. No undercarriage data available.

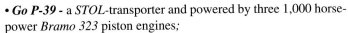

• *Go P-39* - a *STOL*-transporter and powered by three 1,000 horse-power *Bramo 323* piston engines;
• *Go P-40A* - a proposed single piston engine with an asymmetrical planform (similar to the *Blohm & Voss Bv 141*;
• *Go P-40B* - a proposed transport with a wing span of 82 feet, length of 56 feet, and a freight capacity of 540 cubic feet;
• *Go P-45* - a proposed single piston engine transport with a wing span of 78 feet, length of 56 feet, and a freight capacity of 407 cubic feet;
• *Go P-46* - a proposed single engine transport with a 80.4 foot wing span and a length of 50.8 feet;
• *Go P-47* - a proposed transport glider with a 88 foot wing span, 68.5 foot length, and a freight capacity of 1,600 cubic feet. Wing span of 87.9 feet (26.8 meters), length of 64.3 feet (19.6 meters), and a height of 23 feet (7.0 meters);

• *Go P-50* - a transport glider for up to 12 men and one automobile, 65.6 foot wing span, 44.6 foot length, and a freight capacity of 368 cubic feet; wing span of 65.6 feet (20 meters), length of 32.8 feet (10 meters), and a height of 11.2 feet (3.4 meters), or in comparison to the normal *Kalkert Ka-430*: wing span of 73.5 feet (22.4 meters), length of 46.8 feet (14.25 meters), height of 18 feet (5.5 meters).
• *Go P-50.01* - a proposed transport glider;
• *Go P-50.02* - a proposed transport glider with a length of 37.4 feet and a freight capacity of 515 cubic feet;
• *Go P-52* - an all wing project similar to the *Horten Ho 229* and powered by (2x900 kilograms thrust) *Jumo 004* engines;
• *Go P-53* - an all wing project similar to the *Horten Ho 229* and powered by (2x900 kilograms) *Jumo 004* engines;
• *Go P-56* - a tug (*DFS 230/Ju 87*) to be used in combination with a *Fw 190* with glide flight the sailplane *Go P-56*. It is thought to be a tanker and landing on a skid;

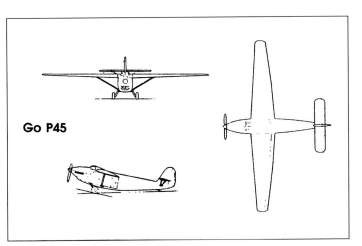

Go P45

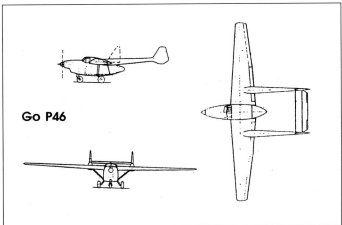

Go P46

The proposed *Gothaer Go P-45.01* transport. This was to have been a conventional high wing monoplane with a strut-braced wing and tailplane. It would have been powered by a single in-line piston engine, fixed two wheel undercarriage, and retractable nose skid.

The proposed Gothaer *Go P-46* transport. A twin boom monoplane with a strut braced high wing and powered by a single *Jumo 211F* piston engine. Its landing gear consisted of a fixed tricycle undercarriage. To facilitate loading and unloading its rear cargo door was hinged at the top.

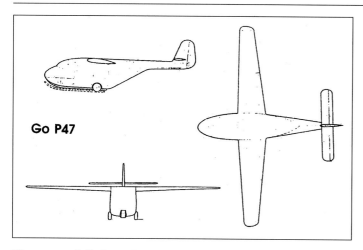

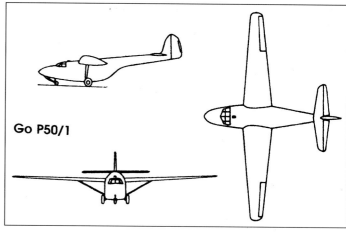

The proposed Gothaer Go P-47 transport glider. It would have been similar in appearance to the Gothaer Go P-50 transport glider, which called for a large square-section fuselage. The undercarriage would have consisted of two wheels and a retractable nose skid.

The proposed *Gothaer Go P-50.01* transport glider. It was supposed to carry up to 12 fully equipped storm troopers.

The individual selected to lead the battle to kill the *Horten Ho 229* was *Dr. Rudolf Göthert*. Amazingly, *Göthert* was not attempting merely to replace the *Horten Ho 229* with a conventional plan form...he claimed that he had something better than the *Horten's* all-wing prototype in mind...a tailless machine, and this is why the *Horten Ho 229* had to die. Not a better all-wing, but a tailless fighter. The academic-only experienced *Dr. Rudolf Göthert* was attempting to beat the *Hortens* at their own game, despite the fact that he had never designed, constructed, or flown even one of his own flying machines. The sum total of his aircraft design work came only from wind tunnel research.

The *Horten* brothers were all-wing and nothing else. So, even though *LFA, DVL, Göthert brothers,* and others associated with them had the academic degrees and the giant *Luftwaffe-* sponsored research centers behind them, the *Horten* brothers were still a team to be reckoned with. They were formidable competitors. After all, the *Horten* brothers had started their aircraft design and development as members of the sailplane community out at the *Rhön/Wasserkuppe* in the early 1930s. This sailplane community was a tight one, and those so-called "*Rhön Red Indians,*" due to their heavily tanned faces from the hot summer sun, fiercely supported one another. *Willy Messerschmitt,* who started his career with the construction of sailplanes (he did not use the *Rhön/Wasserkuppe*), boasted that what he knew about sailplanes...well, he could design one blindfolded. *Reimar Horten* could have said the same. In addition, the *Horten* brothers had designed, built, and test flown a turbojet engine powered aircraft...their *Horten Ho 9 V2.* This significant feat placed them among only a handful of giant German aviation factories who could boast the same: *Messerschmitt AG* with their *Me 262, Arado* with their *Ar 234B* and *Ar 234C,* and *Heinkel AG* with their *Heinkel He 162* and the earlier experimental *Heinkel He 178* and *Heinkel He 280* turbojet powered aircraft. But the *Horten's* feat was even more spectacular. *Messerschmitt, Heinkel,* and *Arado* had hundreds of aircraft designers, a virtual army of aviation engineers, machin-

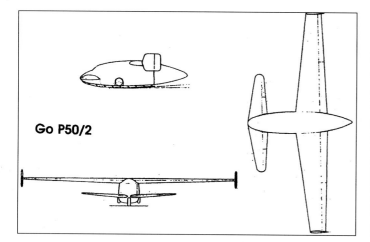

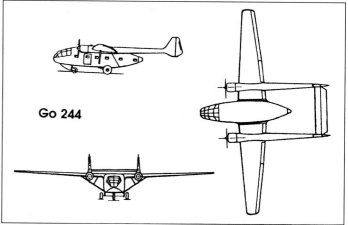

The proposed *Gothaer Go P-50.02* transport glider. This "tail first" monoplane design would have had a long span candard (horizontal stabilizer) and vertical fins with an attached rudder on each wing tip. The last 1/4 of the aft fuselage would have opened upward for loading and unloading cargo.

The proposed *Gothaer Go 244* assault glider...an earlier design.

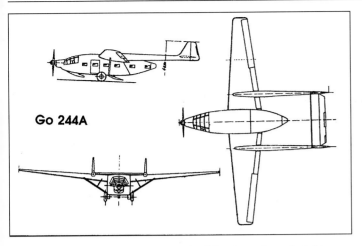

The proposed Gothaer Go 244A assault glider, featuring twin tail booms with its horizontal stabilizer between the booms.

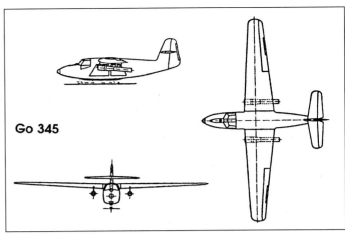

The proposed *Gothaer Go 345* transport/assault glider. A high wing monoplane with a fairly high aspect ratio swept back leading edge wing and a swept forward trailing edge. This gave the wing a fairly pronounced taper. In addition, this flying machine would have had a tall vertical stabilizer with an attached hinged rudder. To facilitate loading, only the pilot's cockpit was hinged to open upwards.

ists, metal fabricators, and so on, all working in modern factories and supported by mountains of financial resources to see their dreams take flight. The *Horten* brothers had a nice shop, a former *Autobahn Maintenance Vehicle Repair Facility (AMVRF)* outside the city of Göttingen given to them by their friend and supporter *Oberst Artur Eschenauer* from the *Luftwaffe Quartermaster General*-Berlin. *Walter* had piloted the legendary public works constructor of the *Third Reich*, General Dr.-Ing. *Fritz Todt* (1891-1942), around coastal France in a *Messerschmitt Bf 108*. *Todt*, the leader of *Organization Todt*, told *Walter* that if he wished he could make full use of the empty *AMVRF* to construct *Reimar's* prototype twin turbojet-powered all-wing fighter. A few days later, *Dr. Todt* sent a memo to *Artur Eschenauer* gifting the use of the *AMVRF* to the *Horten* brothers. Indeed, *Reimar Horten* might have been the first in Germany to have a turbojet powered aircraft in the air (this does not count *Heinkel's* early and short-lived experimental effort with

his *Heinkel He 178*) had the initial *Bramo/BMW 3203* turbojet engines been field ready. Instead, he had to wait and wait to obtain *Jumo 004Bs*, and so his *Horten Ho 9 V2* made its maiden flight on December 18th, 1944. *Messerschmitt, Heinkel,* and *Lippisch,* in their early years, never had the benefit of wind tunnels to test the aerodynamics of their latest design. *Reimar Horten* never did right up to Germany's unconditional surrender on May 8th, 1945. Then again, who did *Hermann Göring* and his aid *Oberst Siegfried Knemeyer* call upon to build the turbojet powered *Amerika Bomber* in late 1944? It was the *Horten* brothers and their all-wing *Amerika Bomber* project, known as the *Horten Ho 18*. So the old timers, such as *Messerschmitt* and *Heinkel,* knew a lot about the passion which burned in *Reimar's* soul, for they too had had that same passion to see their designs take flight, and you could have stood *Reimar* up at the gates of hell and he still would not have backed down. The same could be said for *Messerschmitt, Heinkel,* and *Lippisch*. So,

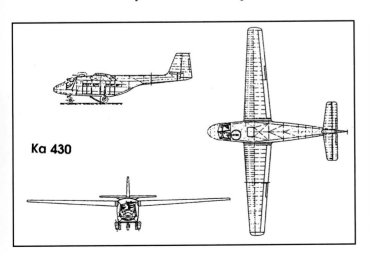

The proposed Kalhert *Ka 430* transport glider.

At one time in the late 1930s there appeared a spark of aircraft design innovation at Gothaer Waggonfabrik. It came from the mind of August Kupper (1905-1937) in the form of the tailless Gothaer Go 147B air observation prototype aircraft of 1936. *Kupper* came to *Gothaer* from *DVL* in 1933. *Kupper's* tailless aircraft designs featured swept back wings with wing tip rudders for directional control.

With the untimely death of *Kupper* in 1937 in the crash of one of his prototype all-wing flying machines, *Gothaer Waggonfabrik* discontinued all development work which *Kupper* had started with his *Gothaer Go 147*. Instead, the firm's design activity concentrated on light aircraft, and its production capacity was turned over to the serial production of war machines, such as the twin motored *Messerschmitt Bf 110* destroyer.

one has to sort of wonder just what was going on in the mind of *Dr. Rudolf Göthert* when he boldly informed the *RLM* that the *Horten Ho 229* which he and his company were instructed to manufacture in series production was just no damn good, and that he, *Dr. Rudolf Göthert*, along with his brother *Dr. Berhard Göthert*, and others from *DVL* and *LFA* were going to give the *RLM* a superior all-wing? *Dr. Rudolf Göthert* was calling his advanced and improved all-wing the *Gothaer Waggonfabrik AG Projekt 60B*, or the *Gothaer Go P.60B*.

Dr. Rudolf Göthert was interrogated between June 28th, 1945, and July 5th, 1945, by members of *General George C. McDonald's* (*AAAFI*) task force with particular reference to the *Gothaer Go P.60B*, his proposed *Horten Ho 229* replacement. The following information is taken from these interrogations, as presented in *Technical Intelligence Report #I-6S* of July 7th, 1945. *Gothaer Waggonfabrik* took the performance of the *Ho 229 V6* and compared it to their *Gothaer Go P.60B*.

History of the *Gothaer Go P.60B*

The *Gothaer Go P.60B* was a twin turbojet powered tailless day fighter proposal dating from September 1944. *Dr. Rudolf Göthert* claimed, unashamedly, that its sole purpose was to displace the approved series production all-wing *Horten Ho 229* prototype by the *Horten* brothers. *Dr. Rudolf Göthert* stated that the *Horten* brothers had obtained a special contract for their *Horten Ho 229* by going directly to the *RLM*. The *Hortens* did not have facilities for serial production to manufacture their *Horten Ho 229*, so the *RLM* had ordered *Gothaer Waggonfabrik* to produce the *Horten Ho 229* in series in August 1944. *Dr. Rudolf Göthert* said that *Gothaer*

Waggonfabrik had nearly completed the *Horten Ho 229 V3*, *Horten Ho 229 V4*, and the center section frame for the *Horten Ho 229 V5*. They had also started fabricating the center section frame for the *Horten Ho 229 V6*, and were working on detailed drawings for the *Horten Ho 229 V7* and *Horten Ho 229 V8*. *Dr. Rudolf Göthert* stated that *Gothaer Waggonfabrik* was reluctant to continue production of the *Horten Ho 229* because they believed they could design a superior all-wing machine. He stated that the only *Horten Ho 229* which had flown (*Ho 9 V2*) crashed on its 2nd flight out of the Orainenburg Air Base, located in the suburbs of northwest of Berlin. (This is not

Gothaer Waggonfabrik helped *Messerschmitt AG* build some of their 6,000 *Bf 110s*, such as this *Bf 110G-4*. *Gothaer* achieved a production high of 120 machines per month in September 1944. Courtesy: *Christian Receveur*.

Gothaer Waggonfabrik also helped out Focke-Wulf *Flugzeugbau* for a time, constructing some of the 4,500 twin motored *Fw 58 "Weihe"* produced. Production started in 1935 as a transport and training aircraft. They had also started a production line for the *Focke-Wulf Fw 152* near war's end and were able only to complete five examples.

Gothaer Waggonfabrik had also entered into a contract with the *RLM* to construct in series the *Focke-Wulf Fw 152*. However, by the first week of April 1945, when the U.S. 3rd Army had secured the Friedrichroda area, *Gothaer* was working on subassemblies but had not delivered any **Fw 152** fighters ready for the front.

entirely correct. Flight testing of the prototype *Horten Ho 229*—the *Horten Ho 9 V2*—began at Oranienburg on December 18th, 1944, and ended on February 18th, 1945, when the machine went into a flat spin with a turbojet engine out during a landing approach, and crashed with *Leutnant Erwin Ziller* still at the controls. However, *Ziller* was thrown out of the all-wing machine when it crashed down on the frozen, snow-covered ground. There was no radio communication between *Ziller* and the ground at the time of its fatal crash, so it is not known why the machine went into a flat spin during its final approach and why *Ziller* never attempted to bail out. Nevertheless, between December 18th, 1944, and February 18th, 1945, it is estimated that the *Horten Ho 9V2* made twelve or more test flights. Although *Erwin Ziller* lists only three test flights in his flight log book, the others were not recorded.)

Dr. *Rudolf Göthert* told his (*AAAFI*) interrogators that overall up to five versions of their *Gothaer Go P.60* were planned:

• *Go P.60A* - an all-wing day fighter prototype;

• *Go P.60B* - a tailless day fighter with its twin turbojet engines mounted above and below the aft center section...the *Horten Ho 229* replacement;

• *Go P.60B-1* - a tailless day fighter similar to the *Go P.60B*, however, with a *Helmuth Walter HWK 509* bi-fuel liquid rocket engine installed internally in the aft center section fuselage;

• *Go P.60B-2* - a tailless day fighter with its twin turbojet engines mounted beneath the aft center section fuselage;

• *Go P.60C* - a tailless night fighter with its twin turbojet engines mounted above and below the aft center section fuselage;

The *Gothaer Go P.60A*

The first model *Gothaer Waggonfabrik* suggested to the *RLM* in replacement of the *Horten Ho 229 V3* was a swept all-wing day fighter. Its two pilots were to lay on individual couches next to each other in the nose of the flying machine, one positioned slightly be-

This post card contains the caption: "Greetings from the flyer city of Gotha. This city is the city of pilots of science and industry: where everyone strives for the beautiful. Gotha and every child notices this. And if your heart has joy in the flyers and the beautiful, then come to Gotha, into the flyers city."

A pen and ink illustration from the *Horten Flugzeuge* featuring their tandem seat *Kampfjäger 8-229*. The drawing notes that it features a wing span of 16.8 meters, a wing surface area of 53.6 square meters, and flight ready weight of 8.5 tons.

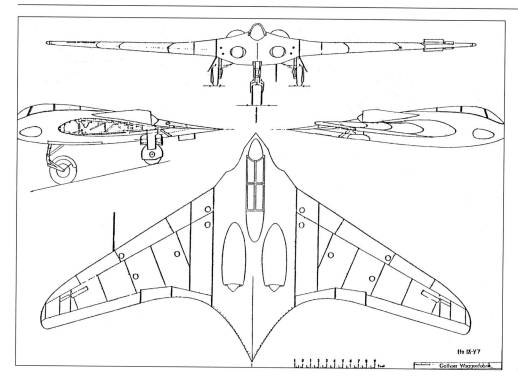

A pen and ink 3-view drawing of the proposed *Ho 229 V7 (8-229)* from the *Horten Flugzeugbau.*

hind the other. Two *BMW 003A* turbojet engines were mounted above and below the aft center section. Its specifications included:

- **wingspan** - 40 feet 8 inches;
- **wing area** - 504 square feet;
- **flying weight** - 16,390 pounds;
- **flying weight overloaded** - 18,800 pounds;
- **fuel capacity** - 475 gallons;
- **speed, maximum** - 596 miles per hour at 23,000 feet altitude;
- **rate of climb at sea level** - 2,750 feet per minute;
- **endurance** - not available;
- **range** - 990 miles at 39,500 feet;

The *Gothaer Go P.60B*

The *Gothaer Go P.60B* was an improved and slightly larger version of the *Gothaer Go P.60A* which was to be powered by twin *Heinkel-Hirth HeS 011* turbojet engines instead of the twin *BMW 003As* in the proposed *Gothaer Go P.60A*. But, unlike the *Gothaer Go P.60A* with its all-wing configuration, the *Gothaer Go P.60B* would have had a tailless configuration. It, too, was to have had a crew of two, both laying in a prone flying position. Design specifications of the *Gothaer Go P.60B* included:

- **wingspan** - 44 feet 4 inches;
- **wing area** - 588 square feet;

A tandem seat Horten Ho 229 night fighter as seen from its port side. It carries the FuG 240 "Berlin 1a" nose mounted radar. Under its port wing can be seen a pair of guided missiles. Scale model and photographed by *Reinhard Roeser.*

• **flying weight** - 22,000 pounds;
• **flying weight** overloaded - not available;
• **fuel capacity** - 915 gallons;
• **speed, maximum** - 608 miles per hour at 16,500 feet;
• **rate of climb at sea level** - 3,740 feet per minute;
• **endurance** - 3 hours and 6 minutes;
• **range** - 1,645 miles at 39,500 feet;

The *Gothaer Go P.60B* version was entered in the *RLM's* January 1945 design competition for a day fighter. Its competition included:

• *Blohm und Voss Bv P.212.03*
• *Focke-Wulf Fw Ta 183*
• *Heinkel He P.1078A*
• *Heinkel He P.1078B*
• *Junkers Ju EF 128*
• *Messerschmitt Me P.1101*

The *Gothaer Go P.60C*

The *Gothaer Go P.60C* was a proposed night fighter version of the *Gothaer Go P.60B* tailless flying machine. It, too, was to have been powered by twin *Heinkel-Hirth HeS 011* turbojet engines. It would have been necessary to enlarge the center section nose to house the internally mounted *FuG 240 "Berlin"* radar scanner (*Spiegel*), and this change would have allowed the crew of three to sit upright (normal position) in the cockpit cabin. The extended nose of the *Gothaer Go P.60C* would have required the addition of two vertical tail surfaces, each with a hinged rudder near each wing tip. In addition, *Dr. Rudolf Göthert* stated that their *Gothaer Go P.60B,* when built in series, would have had the vertical surfaces with an attached hinged rudder, thus giving it a tailless configuration.

The twin turbojet engines for all three *Gothaer Go P.60* versions were to be mounted externally at the rear of the center section: one above and one below in the plane of symmetry. An alternate design with the turbojet engines partially buried side by side in

The incomplete single seat **Horten Ho 229V3** center section discovered by the American 3rd Army at the *Firma Ortlepp*, Friedrichroda. The while metal-like box between the twin *Jumo 004B-2s* was ballast. Head of Air Force intelligence **General George McDonald** had ex-*Gothaer Waggonfabrik* workers attach plywood leading-edge panels before the machine was removed to a *Luftwaffe* collection center for safe keeping. Its sister **Horten Ho 9 V1** flying machine was not so fortunate. Although it too was sent to a *Luftwaffe* collection center, officials apparently decided that the all-wing sailplane was not worth the effort to bring it to the United States.

the undersurface of the center section was proposed for the *Gothaer Go P.60A*. *Dr. Rudolf Göthert* said post war that this arrangement was only to obtain wind tunnel data for application to possible future designs of larger flying wings. In addition, a *Helmuth Walter HWK-509* bi-fuel liquid rocket engine of 4,409 pounds (2,000 kilograms) thrust could have been added to all three *Gothaer Go P.60* versions for use in takeoff and climb. The rocket engine would have

An overhead view of the **Firma Ortlepp, Friedrichsroda,** where the Horten Ho 229 V3 through Horten Ho 229 V6 prototypes, under construction by Go*thaer Waggonfabrik,* were discovered by the American 3rd Army. The building in the background housed the several *Horten Ho 229s.*

The unpainted plywood skinned Horten Ho 229 V3 center section at the Firma Ortlapp as seen from behind.

Two complete outer wings for the *Horten Ho 229 V3* discovered at the *Firma Ortlapp*, Friedrichroda. The fuel was carried completely in its outer wings. Seen inside the wing at its wing root are silver objects...one of several fuel tanks.

• *Arado Ar NJ-1*
• *Arado Ar NJ-2*
• *Blohm und Voss Bv P.215*
• *Dornier Do P.256*
• *Focke-Wulf Fw P.1*
• *Focke-Wulf Fw P.2*

At the conference on night fighters held in March 1945, the *Gothaer Go P.60C* was shown to have the best performance of these aircraft, and *Dr. Rudolf Göthert* believed it would have won had not the war disrupted the competition. On the other hand, the *Gothaer Go P.60B* was not entered into the *RLM's* day fighter competition because the requirements for a day fighter specified that it have only one turbojet engine. *Dr. Rudolf Göthert* claimed that a performance comparison for the performance of the *Gothaer Go P.60B* with the *Horten Ho 229 V6* showed it to be the superior aircraft. For example, the *Gothaer Go P.60A* had slightly less drag than the *Horten Ho 229 V6*, but the *BMW 003A* turbojet engine had slightly less thrust than the twin *Jumo 004B* turbojet engines, and consequently the high speed at sea level was approximately the same. At higher altitudes the high speed of the *Gothaer Go P.60A* was considerably higher, because the critical *Mach* numbers of the *Gothaer Go P.60A* were higher.

The wind tunnel program for the *Gothaer Go P.60* series had been started in aviation research centers throughout Germany. For example, tests on the proper location for the proposed flying machine's twin turbojet engine nacelles had been run at Göttingen. *Dr. Rudolf Göthert* had received some of this wind tunnel data prior to war's end. There was no provision in the models tested to simulate the thrust of the twin turbojet engines, but their inlet velocity ratios were varied from zero to almost one by varying the engines' nose plugs and/or the tail cones. Wind tunnel scale models for the

been installed in the center section. *Dr. Rudolf Göthert* stated that the *Helmuth Walter HWK-509* rocket engine would not be used for high-speed because the speed increase would not be large and would thus be unjustified. All three *Gothaer Go P.60* versions would have had pressure cockpit cabins.

Dr. Rudof Göthert said that the *Gothaer Go P.60C* had been entered in the *RLM's* night fighter competition, which was opened by the *Chef der Technischen Luftrüstung* (Chief of Technical Air Equipment) on February 27th, 1945. In total, seven proposed aircraft, including two other tailless aircraft designs, were presented by the following aviation companies:

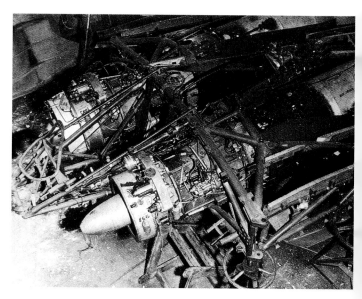

The incomplete aft center section of the single seat Horten Ho 229 V4 discovered at the Firma Ortlapp, Friedrichroda.

The incomplete forward center section of what is thought to be the single seat Horten Ho 229 V5 or V6 found at Friedrichroda. To the left of the photograph can be seen the cockpit wing screen tubular framework. The pilot's cockpit was cramped. Notice the closeness of the twin *Jumo 004 B-2* turbojet engines.

A bare metal framework center section for an unidentified single seat **Horten Ho 229**. It also was discovered at the *Firma Ortlepp*, Freidrichroda.

remainder of the *Gothaer Go P.60B's* wind tunnel program were under construction at war's end. They would have been run in various wind tunnels, including such facilities as:

• **High Speed *DVL* Wind Tunnel - Berlin**
The wind tunnel at the *Deutsche Versuchsanstalt für Luftfahrt* (*DVL*) had a 6.6 foot (2 meter) throat. A complete scale model was to be tested here for its high speed characteristics, as well as a scale model for determining the optimum fillet between the wing and turbojet engine nacelles.

• **Large *LFA* Wind Tunnel - Braunschweig**
The large wind tunnel at the *Luftfahrt Forschungs Anstalt* (*LFA*) had a 26 foot 2 inch (8 meter) throat. The tests in this tunnel were to consist of static longitudinal and directional stability, control surface characteristics, high lift device tests, and general investigation of tip stall characteristics. In addition various wing profiles, control surfaces, and tabs were to be tested.

The center section nose of the **Horten Ho 229 V3** after it had arrived at Wright-Patterson Air Force Base, Dayton, Ohio, in late 1945. Notice the difference between this photograph and the photograph of the same machine when discovered at Friedrichroda. Its center section has been worked on...probably while it was still in Germany.

• **Small *LFA* Wind Tunnel - Braunschweig**
LFA's small wind tunnel had an 8.2 foot (2.5 meter) throat. The damping characteristics in yaw and pitch were to be determined in this tunnel. *Dr. Rudolf Göthert* stated that it was a normal practice to determine these characteristics for proposed aircraft. For models which had a small degree of damping, this model would have been hung with one degree of freedom and the oscillation resulting from a disturbance determined. For models which had a large amount of damping, the resonant frequency would be determined and the energy required to maintain this vibration measured. The damping coefficient would then be calculated from the physical characteristics of model, frequency, amplitude, and energy input.

A port side rear view of the single seat **Horten Ho 229 V3** as seen at Wright-Patterson Air Force Base in late 1945.

A two man *FuG 240 Berlin 1a* radar-equipped *Horten Ho 229 V7* on night maneuvers passing in and out of a group of British *Lancaster* heavy bombers over Germany. Courtesy *Dragon Models*.

Dr.-Ing. Berhard Göthert of DVL.

• **Göttingen University Wind Tunnel**

The tests on the proposed twin turbojet engine installation were to be continued at Göttingen University. Additional wind tunnel scale models for these tests were under construction at war's end. *Gothaer Waggonfabrik* had also planned to make flight tests on the *Horten Ho 229 V3* to obtain data for use on their *Gothaer Go P.60B*. These tests would have included:

— determination of neutral points;
— effects of sweptback on lateral stability;
— effectiveness of various types and forms of rudders, including drag rudders. Some of these tests would have been run on the all-wing twin piston engine *Horten Ho 7*;
— testing various types and forms of landing flaps.

Wing Design

The overall plan form of the *Gothaer Go P.60B* consisted of a center section, which had an NACA airfoil section 0012.5 - 0.850-35 and outer wing panels with airfoil section as follows:
— root section NACA 0012-0.55-50 Y=10 degrees
— tip section NACA 0010-1.1-30* Y=10 degrees

The drawings of the *Gothaer Go P.60B* showed a maximum thickness to be 40% at the tip, but the point of maximum thickness was moved forward to 30%, according to *Dr. Rudolf Göthert*. This was a standard NACA rotation used by the Germans. The root section was a NACA symmetrical profile of 12% chord with its maximum thickness located at the 50% chord, a nose radius of t/c=0.55 (t/c)x2, and it included a trailing edge angle of 10 degrees. The wings had 45 degrees of sweptback, measured to the lines of quarter chord points, and a taper ratio of 0.36.

Dr.-Ing. Rudolf Göthert photographed post war.

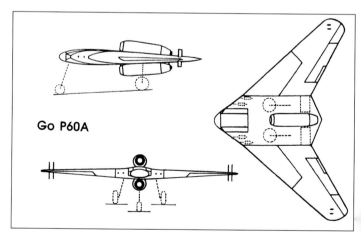

Go P60A

A pen and ink 3-view drawing of *Dr.-Ing. Rudolf Göthert's* first all-wing twin *BMW 003* turbojet engine flying machine for *Gothaer Waggonfabric*...the *Go P.60A-1*. The pilot's configuration would have been a first for a series produced flying machine, in that the two pilots would have flown the machine in a prone position. It would have had a wingspan of 40 feet and 8 inches. Its estimated maximum level forward speed was 596 miles per hour, with a range of 990 miles at 39,500 feet altitude.

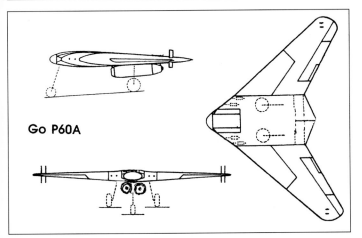

Go P60A

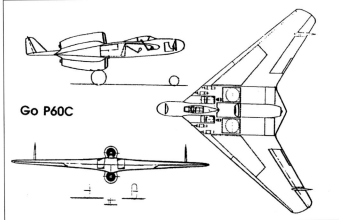

Go P60C

A pen and ink 3-view drawing of Dr.-Ing. Rudolf Göthert's modified Go P.60A-1 known as the Go P.60A-2. The major difference between the two was that in the latter version, *Göthert* removed the upper *BMW 003* turbojet and placed it along side the underside aft *BMW 003*.

This proposed flying machine, the 3rd in the *Go P.60* series, was to have been a night and bad weather fighter carrying three men...all sitting upright. It would have carried the *FuG 240 "Berlin"* parabolic scanning radar in its nose. Performance of the *Go P.60C* was about the same as the *Go P.60B*

A symmetrical profile was chosen for its high speed characteristics, primarily with regard to moments. A zero moment coefficient, or the symmetrical wing, was considered especially desirable at high *Mach* numbers, in view of the considerable troubles with longitudinal trim experienced with other aircraft, and of particular importance on a flying wing. A study had been made for the *Gothaer Go P. 60B* of a wing with the necessary camber, which was found to be only 1% and gave only a 2/3 miles per hour increase in speed. *Dr. Rudolf Göthert* also mentioned ease of construction as an advantage of going with a symmetrical airfoil.

The distribution of the maximum thickness from the 50% chord at the root section to the 30% chord at the tip section was primarily for the improvement of control surfaces. This effect had been observed in wind tunnel tests, and can be seen in *Figure A*:

The effect was primarily to bring about an increase in maximum effectiveness of the control surfaces, which was most important at low speeds, during which large deflections are used, although there may also have been a very slight increase in effectiveness at small deflections. The explanation of this, according to *Dr. Rudolf Göthert,* was based on pressure distribution. The further forward

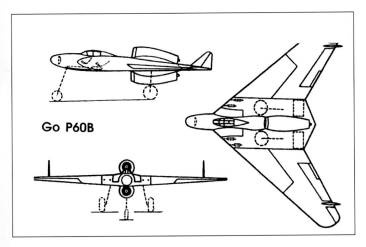

Go P60B

A pen and ink 3-view drawing of Dr.-Ing. Rudolf Göthert's 2nd jet powered flying machine, known as the Go *P.60B*. This proposed flying machine was no longer tailless like the *Go P. 60A-1* or *Go P.60A-2*. Instead, *Göthert* changed the former wing tip "stabilizers" into small vertical stabilizers with hinged rudders. Also, the two pilots were seated in a conventional manner. The Go *P.60B* had an increased wingspan over the *Go P.60A*....44 feet 4 inches. Maximum estimated level speed was 600 miles per hour from its twin *Heinkel-Hirth HeS 011* turbojet engines. Endurance was not as great as the *Go P.60A*. At full thrust the Go *P.60B* could stay in the air for only 33 minutes at 33,000 feet, or 24.9 minutes at 16,500 feet.

The cream of German aerodynamicists strongly opposed the **Horten Ho 229**, but just as strongly supported the *Göthert* twins' proposed tailless versions of their *Gothaer Go P.60B* and *P.60C*. These men included left to right (in front row): *Professor Schlichting* (Experimental Aviation Research Center, Brunswick); *Wolff* (Focke-Wulf); *Dr. Conradis* (Focke-Wulf); *Schomerus* (Messerschmitt AG); and *Husemann* (RLM). Back row left to right: test pilot *Hans Sander* (Focke-Wulf); *Professor Albert Betz* (AVA); *Professor Blenk* (Experimental Aviation Research Center, Brunswick); and *Gotthold Mathias* (Focke-Wulf).

Dr.-Rudolf Göthert's proposed aircraft designs at Gothaer Waggonfabric had full access to the giant wind tunnel at DVL-Berlin-Adlershoff. Shown here at the giant DVL wind tunnel is a Fiesler Fi 103 "Buzz bomb," which terrorized Great Britain.

Gotthold Mathias, aerodynamicist for *Focke-Wulf*. Photographed by the author at Rodondo Beach, California, March 1987. He did not have a very high opinion of the *Horten* brothers' abilities...just like his academic colleagues at *DVL*, *AVA*, and *LEF*.

At high speed, it was *Dr. Rudolf Göthert's* opinion that it would be possible for the airfoil section with maximum thickness at 50% to have better control characteristics. He said that he knew of no systematic wind tunnel tests up to May 1945 which showed the results of his thinking, but he believed the effects would later be shown post war to be approximately as seen in *Figure B*:

Dr. Rudolf Göthert was particularly uncertain of the portion of this curve at low *Mach* numbers. The taper in wing thickness, said *Dr. Rudolf Göthert,* was to be primarily for structural considerations. The sharp nose was used for better high speed characteristics, for example, to get higher critical *Mach* numbers, especially at higher lift coefficients. An illustrative case includes:

the pressure peak, the greater the change in pressure produced by control deflections. Also, there was a secondary effect in that the further back the maximum thickness and corresponding pressure peak, the steeper the adverse pressure gradients on the aft portions of the wing, and the greater the boundary layer effects on the control surfaces.

CL Critical *Mach* Number
Normal Nose Sharp Nose
At CL = 0.1 0.300.82
At CL = 0.4 0.700.80

Post war England gathered up as many of the leading German aerodynamicists and kept them in England for lengthy interrogations. One of the brightest was H*ans Muthopp*...seen 4th from the right in the 2nd row. *Professor Albert Betz* is 2nd from the left in the 1st row.

The explanation of this offered by *Dr. Rudolf Göthert* was that the sharp pressure peak at the nose found at low speed with sharp nosed aircraft was decreased and flattened at high speed, so that it was less than the major (critical) pressure peak further along the airfoil.

Because of the additional pressure on the nose, a higher lift coefficient could be obtained for the same maximum pressure peak. Thus, higher critical *Mach* numbers would be obtained for the same lift coefficient.

Wing root stalling was insured by the extended leading edge flaps on the outer wing, and at high speed, the compressibility shock was also expected first at the root, retaining good flow over the controls at both extremely high and low speeds.

The following summarizes the steps taken to ensure inboard stalling taken on the leading edge when the flaps were retracted:

• distribution of nose radius;
• taper ratio;
• twist (washout);
• thickness;
• the up elevator deflection required to trim at high lift would unload the tip and be beneficial.

With regard to the effects of nose radius on stalling characteristics, *Dr. Rudolf Göthert* believed the percentage nose radius was the significant parameter. With regard to wing thickness, a thin tip with maximum thickness located forward is especially beneficial on a swept back wing because of the cross flow and thickening of the boundary layer at the tip.

Dr. Rudolf Göthert believed that if the above five factors were not sufficient, an automatic slat would be incorporated, in which case the leading edge flap would be dispensed with. The leading edge flap was preferred to the *Handley-Page* automatic slat (*Messerschmitt Bf 109* type) because of lower drag and ease of construction. In the retracted position the automatic slat increases the drag by fixing the boundary layer transition point near the nose. Furthermore, at very high speeds, it has been reported that structural deflection of the trailing edge of the slat takes place and causes additional drag losses. One inch deflection had been reported on the *Messerschmitt Me 262*.

The one degree of wing twist (washout) used on the *Gothaer Go P.60B* was not so much for the purpose of improving the stalling characteristics, as for the primary purpose of reducing the aircraft's pitching moments at the high speed lift coefficient, for example, to enable trim with zero elevator deflection at high speed.

One of the small diameter wind tunnels at DVL-Berlin/Adlershoff. In the wind tunnel's mouth is a scale model of a proposed heavy bomber. The Horten brothers had no access to any of the German wind tunnels because the academic leadership purposely left the two brothers out.

Professor Gunter Bock.

Professor Ludwig Prandtl (center) of Göttingen University, perhaps the most respected college teacher of aerodynamics in Germany, was friendly to the *Horten* brothers. Although he would have allowed them to use the Göttingen University wind tunnel, he was not authorized to do so. *Walter Horten* stands next to *Ludwig Prandtl*, and *Reimar Horten* is on the far right in the photo.

One of the important features of the *Gothaer Go P.60B's* wing was to have been the incorporation of leading edge flaps, a type of high lift device with which the Germans had been experimenting since 1943. There had been considerable general wind tunnel data on this type of device on various wings obtained at Göttingen University, *DVL*-Berlin, and *AVA*-Brunswick which could be found in the reports of these institutions. The type of flap to be used on the *Gothaer Go P. 60B* was sketched by *Dr. Rudolf Göthert* and is shown in *Figure C*:

Dr. Rudolf Göthert claimed that the leading edge flap's aerodynamics essentially increased the radius of curvature of the nose, as well as increasing the effective camber, and thus postponed the separation of the flow on the upper airfoil surfaces, so that the stalling characteristics were much like a simple slot as shown in *Figure D*:

The moment characteristics of the leading edge flap had another advantage according to *Dr. Rudolf Göthert*. This advantage was that there was no change in either trim or slope of the moments up in the high lift range. In addition, after the section had stalled the changes in moments were much less severe than with a simple airfoil. This was explained by *Dr. Rudolf Göthert* as resulting from the large effective chord with consequent forward shift of the center of pressure. There was some change in trim at low angles of attack, but the leading edge flap would normally be retracted under those type of flight conditions. See *Figure E*:

Dr. Rudolf Göthert told his (*AAAFI*) interrogators that the purpose of the leading edge flaps on the *Gorthaer Go P.60B* was not particularly to increase the maximum lift of the airplane, but because the low wing loading gave good stalling characteristics and insured good control characteristics at the stall. For sharp nosed sections, an increase in section maximum lift coefficient on the *Gothaer Go P.60B* varied from 0.15 at the tip to 0.30 at the root, and the increase in airplane maximum lift coefficient was expected to be 0.9 from 0.8. The advantages to be gained with this type of flap were, of course, greater with airfoil sections having sharper noses.

When questioned as to whether the leading edge flows had any effect on the lateral stability (rolling moments due to yaw), *Dr.*

A unique person in the highly regimented Nazi controlled Germany, Hans Multhopp (right) was friendly to the Horten brothers, especially Reimar. Ludwig Pra*ndtl* called *Multhopp* the most gifted student of aerodynamics he had ever taught. This included *Professor von Karman*, also a former student of *Prandtl. Multhopp* encouraged *Reimar Horten* to continue on with his all-wing work, telling him that his work "could be (correct), it could be." But M*ul*thopp was unable to get *Reimar's* all-wings into any of the several wind tunnels throughout Germany. *Multhopp's* assistant, in Germany and later in the United States, *Martin Winter*, is seen on the left.

Hans Multhopp's late war ultra modern turbojet powered day fighter design...his Fw Ta 183, as seen from its starboard side. Scale model and photographed by Günter Sengfelder.

A ground level nose starboard view of *Hans Multhopp's* high advanced day fighter...the *Fw Ta 183*. Scale model and photographed by *Günter Sengfelder*.

A pen and ink illustration of Dr.-Ing. Rudolf Göthert's Horten Ho 229 killer. His all-wing *Gothaer Waggonfabric Go P.60A,* powered by twin BMW *003* turbojet engines, was estimated to reach a top speed of 596 miles per hour in level flight at 23,000 feet altitude.

A port side view of the *BMW 003A-1* turbojet engine. This is pretty much the position the upper *BMW A-1* would have taken as installed on *Göthert's Go P.60A* all-wing flying machine.

Rudolf Göthert told his (*AAAFI*) interrogators post war that he was not certain, although he thought the effect on the *Gothaer Go P.60B* would be small. He mentioned that in some unpublished wind tunnel test results from Prague (Czeckoslovia) by the *DVL* with *Arado Flugzeugbau* showed some small effects to be gained.

Lateral Stability

Lateral stability was recognized by *Dr. Rudolf Göthert*, according to his (*AAAFI*) interrogators, as one of the most difficult problems of an all-wing flying machine. He did not consider any solution satisfactory at the end of the war. The problem, according to *Dr. Rudolf Göthert*, was to have sufficient lateral stability (rolling moment due to yaw) at high speed (low lift coefficients) when there is little contribution from the sweep of the wing, and most of the sta-

The BMW 003A-1 turbojet engine of the type Göthert would have placed in this Gothaer Go P.60 A day fighter and seen from its nose starboard side. It was capable of producing 1,320 pounds of thrust at 560 miles per hour, or 1,500 pounds static thrust at sea level.

bility had to be obtained from dihedral. At the other end of the speed range (at high lift coefficients) the sweep of the wings contributed a large measure of lateral stability, which was in addition to the dihedral effect needed at high speed, resulting in an excessive amount of lateral stability in low speed conditions.

The *Gothaer Go P.60B* was to have had 1.0 degree of dihedral, which would have given it a satisfactory lateral stability at high speed. However, *Dr. Rudolf Göthert* stated that he was not sure that the low speed lateral stability would be satisfactory, and if flight tests would indicate such to be the case. Nonetheless, he was thinking of using an automatic mechanism to artificially produce the lateral stability. This mechanism would have been in the form of an aileron actuating mechanism using a yaw vane as a signal, that is, a disturbance in yaw would cause the yaw vane to transmit a signal that would electrically actuate the ailerons. This activation would produce a rolling moment in the desired direction. *Dr. Rudolf Göthert* stated that he was not interested in the mechanics of the device, but only in the fact that could actuate the ailerons.

Dr. Rudolf Göthert told his (*AAAFI*) interrogators that the *Blohm und Voss Bv 141* asymmetrical flying machine had the same type of electrical aileron device that produced artificial lateral stability. In the *Blohm und Voss Bv 141* this was achieved by two small vanes, both with different moment arms, mounted on the lower surface of the wing, one pair ahead of each aileron. See *Figure E.* These arms were linked to the aileron mechanism such that the aerodynamic forces on the vanes contributed to the aileron hinge moments as soon as the ailerons were deflected from a neutral position.

The vanes were used for the dual purpose of reducing aileron control forces and producing apparent lateral stability. *Dr. Rudolf Göthert* stated that he would have used such a device only as a last resort, because it was in the nature of a "makeshift" device for correcting an existing unsatisfactory design.

In the preliminary design stages of the *Gothaer Go P.60B, Dr. Rudolf Göthert* stated that he would attack the lateral stability prob-

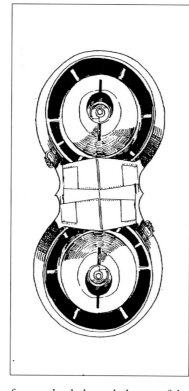

A pen and ink illustration of the proposed "over and under" BMW 003A-1 turbojet engine placement on the aft center section of Göthert's Go P.60A all-wing day fighter as seen from behind the aircraft.

Directional Stability and Control

The original *Gothaer Go P.60* designs...the *P.60A* and *P.60B*...had no vertical tails, and the static directional stability was considered satisfactory due to the absence of protrusions of any sort on the nose and the beneficial vertical fix effect of the engine nacelles along the rearward portion of the wing.

However, on the *Gothaer Go P.60C*, the night fighter rear equipment included a 3 foot diameter reflector..."*Spiegel*." This device required an enlarged and extended nose section, as well as a canopy for the upright-sitting pilot and copilot. To make up for these unfavorable influences on directional stability, *Dr. Rudolf Göthert* would have added vertical fins to this night fighting version. They would have been placed outboard on the wings between the two control surfaces.

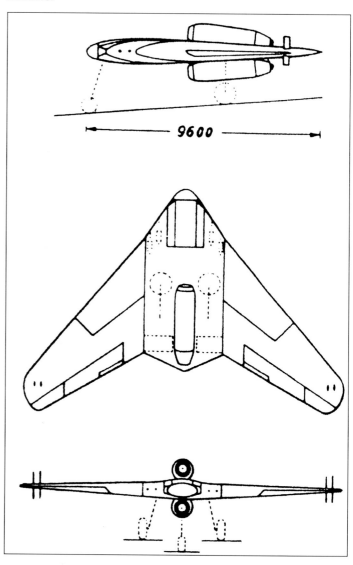

9600

A pen and ink 3-view drawing of the *Gothaer Go P.60A* by *Rudolf Göthert*. The wing was to have 45 degrees of sweep back. About the time *Gothaer Waggonfabrik* obtained the contract to construct the *Ho 229* in series, *Göthert* himself had been working on two all-wing designs within known as the *Gothaer Go P.52* and *Go P.53*. Friends of the *Horten* brothers claimed that *Göthert* combined the all-wing features of the *Ho 229* along with his *Go P.52* and *Go P.53*, and the result was the *Gothaer Go P.60* series of flying machines.

lem by minimizing the effects of sweep back through the use of the proper amount of wing tip design taper. The effect of wing tip shape would be a small one, but optimum design would be such to minimize the amount of the tip area at angles of yaw as illustrated in *Figure F*.

Dr. Rudolf Göthert pointed out to his (*AAAFI*) interrogators that, other than in landing conditions, the flying wing would normally be flying at low lift coefficients because of its low wing loading. This was a result of the physical limitation on the density of loading of all-wing aircraft. It was for this reason *Dr. Rudolf Göthert* had used enough dihedral to give satisfactory flying qualities in most of the normal and high speed range, and it was only in landing that he expected unsatisfactory lateral stability.

The *BMW 003A-1* engine cowling would have opened similar to this *BMW A-1* found on the *Heinkel He 162*.

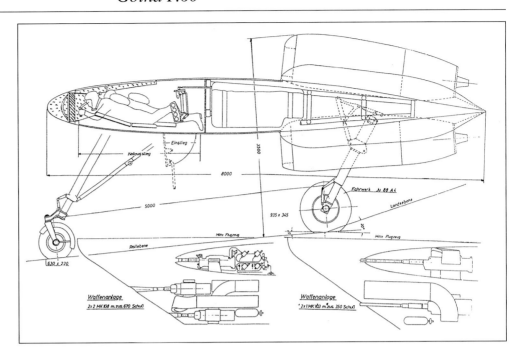

A pen and ink port side view of the all-wing *Gothaer Go P.60A* featuring the pilot's prone piloting position, nose wheel gear, main wheel landing gear, and armament.

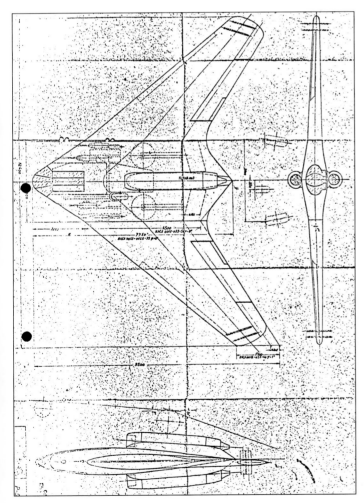

A poor quality pen and ink 3-view drawing of the all-wing *Gothaer Go P.60A* from the *Gothaer Waggonfabrik*. However, included in this drawing are the flying machine's cannon locations and nose wheel retracted area.

The original directional control device on the *Gothaer Go P.60 A* and *Gothaer Go P.60B* versions was a narrow chord airfoil at a pre-set angle of attack, placed near the tip of the wing, and hinged about a span wise axis. It would lie completely enclosed when not in use, and would produce a yawing moment proportional to its deflections. See *Figure G.*

It was pointed out that with the pre-set angle of attack of 15 degrees and the stalling angle of the section 20 degrees, the device would be stalled if the airplane were yawed more than 5 degrees. After thinking this over, *Dr. Rudolf Göthert* admitted to his (*AAAFI*) interrogators that it was a serious drawback to this device. He later admitted that this type of directional control device would not be retained on future designs, but that it had been adopted largely for appearance sake (a cleaner flying wing) on the first models. *Dr. Rudolf Göthert* expected to use the *Gothaer Go P.60C* type of directional control on all models...including the *Gothaer Go P.60A* and *Gothaer Go P.60B*.

The *Gothaer Go P.60C* was to have vertical fins, necessary for static stability as mentioned before, and these were to have conventional rudders attached. These vertical fins were to be placed inboard from the tip for several reasons. One was to use the fin to straighten out the cross flow on the aft portion of the sweptback wing, that is, to prevent the boundary layer from piling up on the outboard control surface, and therefore retaining better control surface characteristics for it, since this was the one supplying the predominant portion of control "feel." The inboard location of the fin would also have less unfavorable effect on the damping roll, that is, enabling a greater rolling velocity and at the same time reducing the structuring problem of high loads on the fins, which would occur in rolling maneuvers.

The directional control and stability problem was considerably alleviated by the lack of any thrust asymmetry with partial engine operation due to the power quotient locations, and the large amount

An MK 108 30mm cannon of the type to be installed in the Gothaer Go P.60 series. At the top of the photo is legendary aviation historian *Alfred Price*.

of sweep back necessary for high speed provides a large moment arm for the vertical tail surfaces. The dynamic directional stability problem caused *Dr. Rudolf Göthert* some concern as to the low directional damping of a flying wing. The directional oscillation was particularly troublesome at high speeds when a steady gun platform was desired. If this problem existed, the two possible solutions were to increase the period of oscillation (*Dr. Rudolf Göthert* mentioned approximately 15 seconds for the *Gothaer Go P.60B*), thereby allowing the pilot to use his controls to dampen it out, or to increase the inherent damping of the airplane. Since increasing the period requires changes in moment of inertia, *Dr. Rudolf Göthert* agreed that the better approach to the problem was to increase the directional damping, and he regarded the vertical fin of the *Gothaer Go P.60C* as important for this reason. He also hoped to get some contribution from the engine nacelles, and the engine-wing fillet had been extended aft somewhat for this reason. In the case of the *Gothaer Go P.60A* and *Gothaer Go P.60B*, the directional control devices could be left in the extended position at high speed for increased damping.

Another type of directional control, a "drag rudder," was being considered. It would have consisted of surfaces extending span wise out of the wing tips. They were to be wind tunnel tested, and had been used by the *Horten* brothers. *Dr. Rudolf Göthert* told his (*AAAFI*) interrogators that he had not spoken to the *Hortens* and had not seen any test results, however, he was not very enthusiastic about this type of device, although it was believed to be superior to the spoiling type of directional control that the *Hortens* had installed on their twin *Jumo 004B* powered *Horten Ho 9 V2* prototype fighter.

Longitudinal Stability and Control

Dr. Rudolf Göthert told his (*AAAFI*) interrogators that he considered longitudinal stability and control of flying wings at high speed could be minimized by low aspect ratio and high taper ratios. At high *Mach* numbers, he emphasized the existence of two stability

A starboard view of an *MK 108 30mm* cannon proposed for the *Gothaer Go P.60*. At the base of the barrel can be seen a single *30mm* shell.

A port side view of an *MK 108 30mm* cannon.

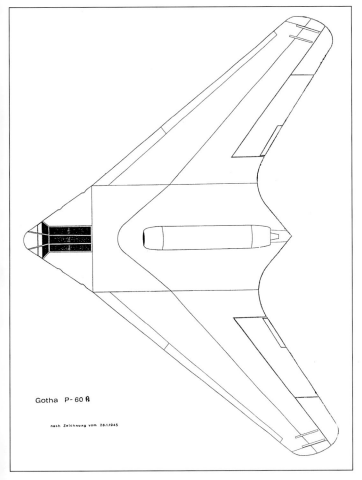

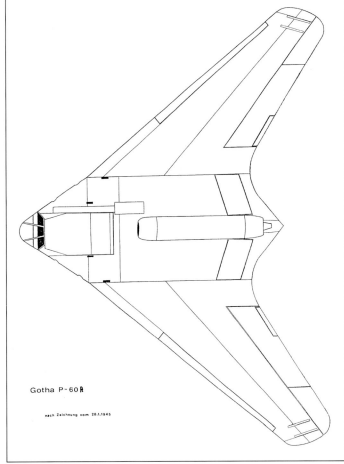

A pen and ink drawing from *Gothaer Waggonfabrik* dated January 28th, 1945, and featuring the all-wing *Gothaer Go P.60A's* upper side and control surfaces along its trailing edge.

A pen and ink drawing from *Gothaer Waggonfabrik* dated January 28th, 1945, and featuring the all-wing *Gothaer Go P.60A's* under side and its tricycle landing gear. Notice that the nose wheel/gear is off center to port to allow more room in the cockpit floor for the prone crewmen.

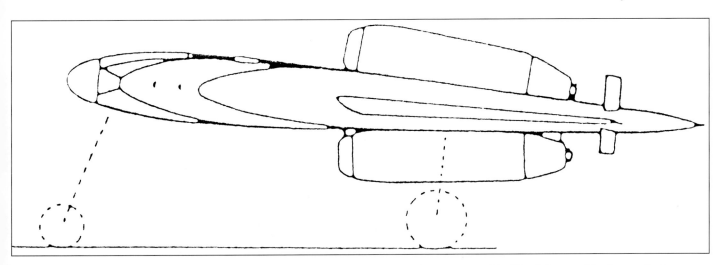

A pen and ink hand drawn illustration featuring the Gothaer Go P-60A as seen from its port side. The flying machine's unorthodox prone piloting position was a concern to the *RLM* planners due to the pilot's inability to see what was coming at him from behind. *Rudolf Göthert* believed that this concern was outweighed by the pilot's need for minimal aerodynamic drag in achieving high forward speed...besides, no enemy could sneak up on this machine.

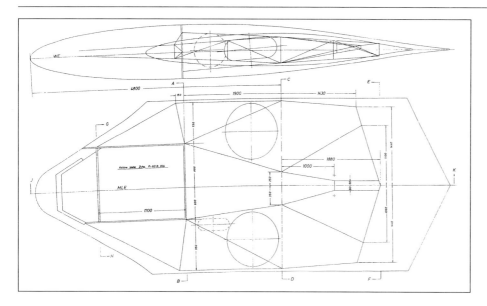

Pen and ink illustrations featuring the center section of the Gothaer Go P.60A: (top) port side center section profile and (bottom) overall top view. The center section would have been constructed similar to the *Horten Ho 229*...metal tubular pipe and covered with aircraft grade plywood sheeting.

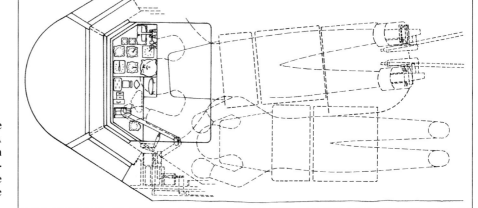

A pen and ink drawing illustrating the prone piloting position for the two Gothaer Go P.60A crewmen. The cockpit cabin would have been pressurized and fully flared into the surrounding nose contour. To provide additional space within the cockpit, the nose wheel would have been offset to port.

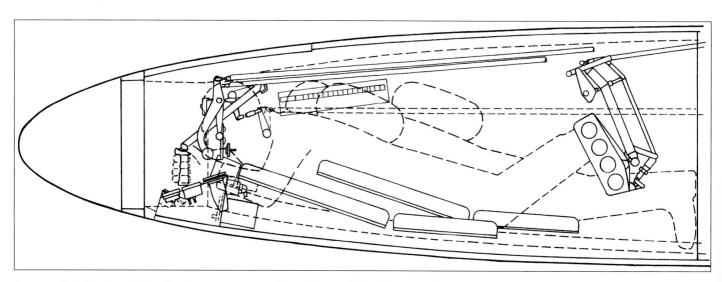

A pen and ink drawing illustrating the center section of the Gothaer Go P-60A as seen from its port side. The dotted lines represent the two man crew's flight positions. Rudder control was achieved via hanging foot pedals. The "elevons" (elevators and ailerons combined into one control surface) were to be activated via hanging levers. There was no conventional "control stick" planned for the *Gothaer Go P.60A*.

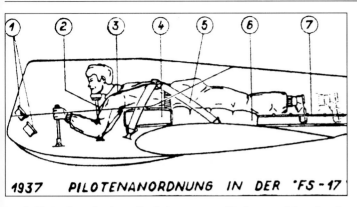

Rudolf Göthert had studied the prone piloting position in the *Flugtechnische Fachgruppe*-Stuttgart FS 17 from 1937, as seen in this pen and ink drawing. He liked the advantages prone piloting presented over conventional upright siting.

Felix Kracht, chief of design at DFS-Ainring, designed his "Rheinland" sailplane for prone flying. The Rheinland is seen from its port side. He would later feature prone piloting in his DFS 228 V2 Walter HWK-509 bi-fuel rocket powered reconnaissance prototype.

criteria, both of which were significant. One was the usual low speed criterion of $2Cm/3CL$ at constant *Mach* numbers, and the other was the complete derivative dCm/dCL with varying *Mach* numbers. The first measure of static stability $2Cm/2Cg$ (with "*m*" being constant), as is well known, varied with *Mach* number, such that the neutral point first moved back until the critical *Mach* number was reached, and then jumped forward rapidly. By going to new aspect ratio and high taper ratio, this effect could be minimized, and *Dr. Rudolf Göthert* had achieved the following for the *Gothaer Go P.60B*. See *Figure H*:

Since most flight maneuvers at high speed involved a change in *Mach* number whenever "*CL*" was varied, the only true indication of stability in these cases was one which took into account the variable *Mach* number. It could be obtained from wind tunnel data by cross plotting against *Mach* number for equilibrium conditions. See *Figure I*:

From this, *Dr. Rudolf Göthert* said it can be seen that even though the neutral point may be moving forward, the stability dCm/dCL "*m*" is constant. Maybe increasing up to the critical *Mach* number. Likewise, past the critical *Mach* number, although the constant *Mach* number neutral point moved back, the stability with variable *Mach* number would be decreasing rapidly.

According to *Dr. Rudolf Göthert, Luftwaffe* flying quality requirements were not concerned with stability in this region so much as the controllability, that is, instability was allowed, but the high speed limit on the *Mach* number was usually the controllability limit. The plot shown above was typical; there was an uncontrollable nose down pitching moment at high speed, as experienced in flight testing on several *Luftwaffe* aircraft. Use of trim tabs at high speed was not recommended because local shock waves formed on the tab. The stability criterion of $2Cm/2CL$ at constant *Mach* number still had significance, in that bumps and gusts affecting the air-

The center section of the prone piloted *Horten Ho 4B* high performance all-wing glider as seen from its port side. This machine had a laminar-flow wing, however, it did not work out as effectively on this sailplane as *Reimar Horten* had planned. Only one example was constructed, and both the test pilot and the machine were lost in a test flight.

Horten Flugzeugbau test pilot *Heinz Scheidhauer* is shown positioning himself in the prone flying position in a *Horten Ho 4A*. We see pilot *Scheidhauer* from the port side.

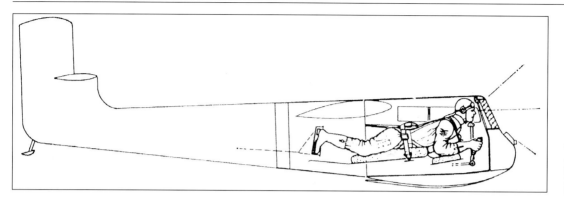

The *Berlin B-9* also had prone piloting, as seen in this pen and ink drawing featuring the B-9's starboard side.

plane while traveling at high speed effectively changed the *DL* while at constant velocity and *Mach* number, and the restoring moments set up in such a case were dependent on the old low speed criterion.

It was suggested to *Dr. Rudolf Göthert* by his (*AAAFI*) interrogators that a high speed controllability limit was not necessarily due to high control forces, and in fact there might be reversal of control forces in that region. This was attributed to overbalance of controls, especially of the sharp-nosed *Frise* type, when a shock wave might form on the nose. In such a case, the stick forces would vary as follows as shown in *Figure J*:

Dr. Rudolf Göthert's response was that German aerodynamicists had done considerable testing of control surfaces at high *Mach* numbers, and the results could be found in reports published by the *Lilienthal Gesellschaft*. These results, *Dr. Rudolf Göthert* claimed, indicated a rapid loss in elevator effectiveness at high *Mach* numbers, eventually passing through zero and becoming reversed. Tests had not been completed in the range of reversed effectiveness, but he believed that the control effectiveness would regain its normal sense at a still higher *Mach* number. *Dr. Rudolf Göthert* sketched the following curve for the (*AAAFI*) interrogators as shown in *Figure K*:

The explanation of this effect was based on the formation of a local shock wave on one side of the control surface, with pressure distribution such that the effectiveness was the reverse of normal. This might also cause a reversal of control forces. At a somewhat high *Mach* number another shock wave might form on the other side of the surface, which was the basis of *Dr. Rudolf Göthert's* belief that the effectiveness returned to its normal sense. The rever-

sal point of high speed static stability depended on the airplane's critical *Mach* number (predominately the wing), reversal point of control effectiveness on the control surface's critical *Mach* number, and the point of control overbalance dependent on the nose balance.

The *Gothaer Go P.60B* was to have an actual center-of-gravity travel of 3 percent. A large center-of-gravity travel (as much as 10%) would be permissible in the middle speed range. *Dr. Rudolf Göthert*, according to his (*AAAFI*) interrogators, seemed willing to push the permissible center-of-gravity further forward, if necessary, at the expense of higher landing speeds, that is, the reduction in the maximum obtainable with full rudder.

Control Surfaces

The *Gothaer Go P.60B* would have used "*elevons*" for both elevator and aileron control, but split into pairs of surfaces. Only the outer set was to be used at high speed, while both sets were to be used at lower speeds. Also, the outer set were to be directly linked to the pilot's control, whereas the inner set was actuated by a servo-tab and the on-board force transmitted from the pilot's control was to be the tab itself, so the inboard *elevon* thus floated freely. The ratio of tab deflection to outer *elevon* deflection was fixed (a non-linear relationship), except for the disconnect at high speed. The relative apportionment of *elevon* deflections to the elevator and aileron function is shown in *Figure L*.

According to *Dr. Rudolf Göthert*, it could be seen that the available aileron deflection was reduced at high speed. This was to be with a constant amount of travel of the pilot's control, so that the

A test pilot is entering the twin motored single-seat *Berlin B-9* with its prone piloting arrangement.

The Henschel Hs 132 prototype dive bomber of 1945 also featured prone piloting. Although no Hs 132s had been completed prior to war's end and the Soviet's obtained what did exist, this illustration from the late *Gerd Neumann* shows how the flying machine might have looked were it completed.

mechanical advantage of the ailerons increased at high speed. The aileron deflections were kept symmetrical (no differential) to prevent a change in longitudinal trim accompanying the use of lateral control.

The outboard elevon was 20% of the wing chord, and the inboard only 15% of the wing chord. The larger percent chord of the outboard *elevon*, which was to be the only control used at high speed, was chosen for the purpose of reducing the severity of the detrimental compressibility effects on control effectiveness at high *Mach* numbers, just below the effectiveness reversal point. This should be clear, said *Dr. Rudolf Göthert* to *General George C. McDonald's* (*AAAFI*) interrogators, especially from the sketch which he drew of elevator effectiveness versus *Mach* number for several chord sizes previously presented.

The *Gothaer Go P.60B's* elevon contours were to be a rather blunt-nosed *Frise* type, which *Dr. Göthert* said had been obtained empirically as the result of much wind tunnel experience (as a member of the *Sonderausschusses Windkanale*, he was a specialist with this group), and were said to have the best possible effectiveness, hinge moments, and drag characteristics. The amount of aerodynamic balance had been tentatively set at 26% (ratio of balance ahead of hinge line chord aft of hinge line) for the outboard control and 25% for the inboard control. The lesser balance was used on the outboard control because it would yield the predominant portion of the pilot's control forces, and the more-linear control characteristics obtainable with smaller amounts of balance were desired. The inboard control surface was to be equipped with a servo tab (*Flettner*, also known as a 100% balancing tab), which was also to be used for longitudinal and lateral trimming. The tab itself would have had a 26% aerodynamic nose balance, giving increased tab

The DFS 228 V2 in flight high over the Cliffs of Dover. Artwork by Ferguson.

effectiveness as well as minimizing tab control forces, which *Dr. Rudolf Göthert* claimed was a common practice in Germany. The tentative balance figures produced were to be subject to revision pending wind tunnel tests, with a possible variation of 1% for the tab and outer control surface, but 2% for the inboard control surface.

Dr. Rudolf Göthert admitted that non-linear characteristics were to be expected from the tab and *elevon* combination, and estimated approximately 30% less in effectiveness at high angles of attack caused by the thickening of the boundary layer at the trailing edge.

In the design of control surfaces for supersonic aircraft, it was mentioned that the hinge line should be considerably aft, probably 50% of the overall control surface chord, because of the shifting of the aerodynamic center to the rear at high *Mach* numbers.

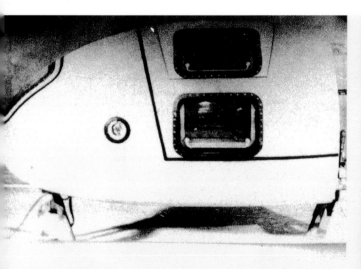

A direct port side view of the cockpit cabin for the *DFS 228 V2*. Rolf Mödel constructed this cockpit cabin out of wood according to *Felix Kracht's* design specifications. There are several changes from the *V1*. The fuselage side windows have been moved higher up on the fuselage wall. The pilot's entry and egress hatch extends down around the fuselage sides plus it carries a window on each side of the fuselage. In the nose we see that the round port hole on the *V1's* cockpit cabin has been changed to a more squarish plexiglass window. Finally, the swordfish emblem is more professional-looking due to being placed inside a circle.

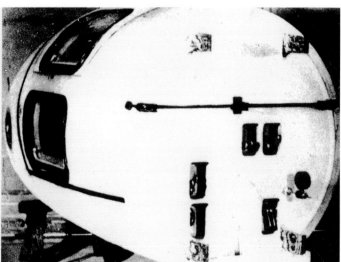

The *V2's* cockpit cabin as viewed from its port rear side. The two metal tabs seen protruding on the rear wall of the cabin both top and bottom, are its attachment points to the center fuselage. Quick, disconnect connections can been seen about center on the fuselage rear wall.

Felix Kracht-designed Walter HWK-509 bi-fuel liquid rocket powered DFS 346 with its prone piloting position. This machine is shown in the USSR in the late 1950s featuring its port side. Soviet aerodynamicists added boundary layer fences to the upper surfaces of both wings.

The starboard side of the prone piloted *DFS 346* attached to the starboard wing of an interned American *Boeing B-29* in the USSR circa 1950.

Performance Estimates of the *Gothaer Go P.60B* Fighter

The performance of the *Gothaer Go P.60B*, furnished by *Dr. Rudolf Göthert*, is based on a *Heinkel-Hirth HeS 011* turbojet engine generating 2,866 pounds (1,300 kilograms) of static thrust. Drag numbers for the *Gothaer Go P.60B* were estimated at 0.510, while the *Gothaer Go P.60C's* drag was estimated at 0.551.

The drag estimate for the *Gothaer Go P.60C* by *DVL*, who calculated the performance of all the night fighter projects entered in the *RLM's* competition, was slightly higher than the other entrees. The critical *Mach* number of the wing alone was 0.835, and the addition of the engines lowered it to 0.82. For drag purposes the German definition of critical *Mach* number was the *Mach* numbers at which the drag had increased by 20 percent. The critical *Mach* number and the increase of drag with *Mach* number had been taken from previous wind tunnel tests of various wings considering aspect ratio, sweep back, and profile section of the *Gothaer Go P.60B* and the *Gothaer Go P.60C*. However, *Dr. Rudolf Göthert* was unable to provide any specific references to these wind tunnel drag tests.

Engine Nacelle Location

Dr. Rudolf Göthert stated to his (*AAAFI*) interrogators that the *BMW 003A* turbojet engines were placed at their present location due to the following reasons:

- heat from the twin *BMW 003A* turbojet engines would be far removed from the wing;
- the critical *Mach* number of the flying machine would be increased by placing the twin *BMW 003A* turbojet engines to the rear of the center section;
- overall drag to the flying machine would be lower with its twin *BMW 003A* turbojet engines located aft on the center section than with them buried in the center section and air coming from leading edge ducts; (*AAAFI*) interrogators questioned this conclusion, stating to *Dr. Rudolf Göthert* that any leading edge drag may have been high due to less than optimum leading edge duct design;
- directional stability would be increased by the aft location of the twin turbojet *BMW 003A* engines;
- external turbojet engine location would allow for better access for maintenance;

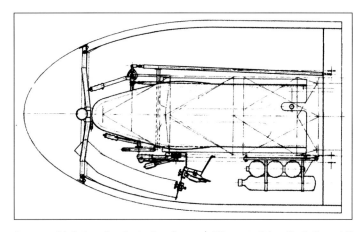

A pen and ink drawing featuring the overall layout of the pilot's "couch" in the cockpit of the DFS 346. Engineering drawing by *Felix Kracht-DFS-Ainring*.

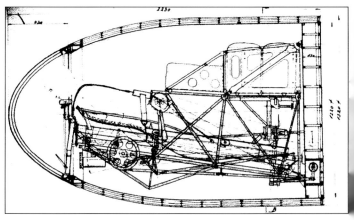

A pen and ink drawing featuring the port side layout of the pilot's "couch or bed" in the cockpit of the DFS 346. The overall outline of the pilot is barely visible. Engineering drawing by Felix Kracht - DFS-Ainring.

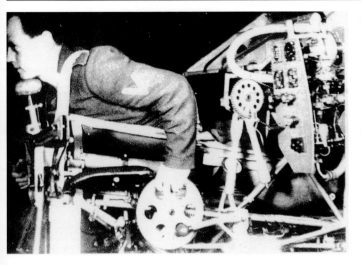

Rolf Mödel of DFS-Ainring shown testing the fit and comfort of the pilot's "couch" or bed" in the cockpit of the *DFS 346. Mödel* constructed the cockpit out of wood from *Felix Kracht's*—his boss'—engineering drawings.

Rolf Mödel of DFS-Ainring seen in the cockpit (port side) of the DFS 346.

Pilot's Compartment

The pilot's compartment in the *Gothaer Go P.60B* was flush with the surrounding center section and featured a standard cockpit cabin, with the pilots seated in the conventional upright position. The type and movement of the controls in both versions was normal, with the exception in the *Gothaer Go P.60B* being that its rudder pedals and control stick were of the hanging type. In addition, there was a control which would have separated the tab-control mechanism from the control stick.

Dr. *Rudolf Göthert* stated to his (*AAAFI*) interrogators that *DVL* had built an aircraft flown in the prone position known as the *Ber-lin B-9* for the purpose of conducting research on prone-pilot cockpits with regard to pilot vision, instrument layout, and ability of the pilot to remain in the cockpit for extended periods. The *Gothaer Go P.60A* and *Gothaer Go P.60B* cockpit designs were based on discussions between *DVL* and *Dr. Rudolf Göthert*. He believed that no extreme discomfort would be encountered by the pilot in a flight up to three hours in length. For longer flights, *Dr. Rudolf Göthert* believed that it would be necessary to have two pilots so that they could alternate flying the machine. With regard to vision, *Dr. Rudolf Göthert* stated that the main difficulty would be a pilot's vision to the rear, however, because of the *Gothaer Go P.60B's* high speed poor rear vision was not considered to be a serious drawback.

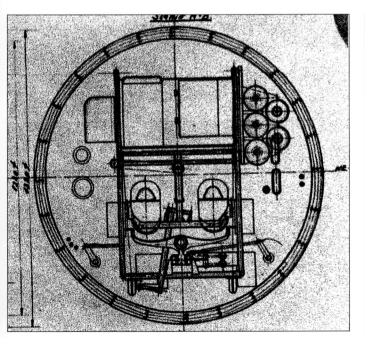

A poor quality pen and ink drawing of the *DFS 346's* detachable prone pilot position cockpit as seen from the rear.

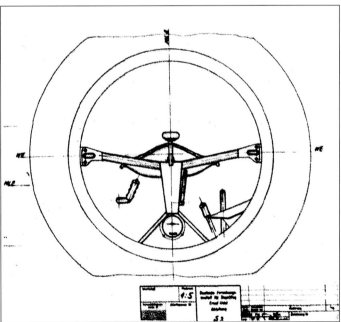

A pen and ink drawing of the DFS 346's cockpit as seen from the front. The oval item in the center of the cockpit's windscreen is the pilot's chin rest.

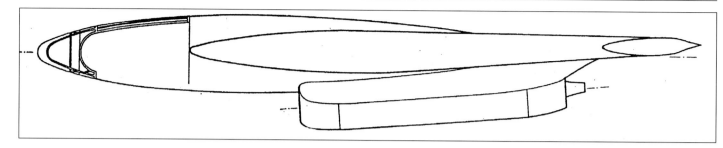

Dr. Ruldolf Göthert's extensive wind tunnel studies on his Gothaer Go P.60A planform suggested that repositioning the twin BMW 003s and placing both under the center section would be advantageous. This pen and ink drawing from Gothaer Waggonfabrik and featuring its port side shows how clean the center section would be with its twin BMW 003s side by side aft under the center section.

Comparison Between the *Horten Ho 229 V6* and the *Gothaer Go P.60B*

Gothaer Waggonfabrik AG personnel wrote this comparison of the *Horten Ho 229 V6* and their own *Gothaer Go P.60B* for the *RLM*. It is dated January 27th, 1945. Its purpose was to show the superiority of the *Gothaer Go P.60B* over the *Horten Ho 229 V6*. The author has translated it from the German.

Arrangement Of Propulsion Units:
• *Horten Ho 229 V6* - Two *Jumo 004B* turbojet engines buried in the center section and straddling the cockpit. Air intakes are located in the center section's leading edge, and exhaust exits out over the center section's tailing edge.
• *Gothaer Go P.60B* - Two *BMW 003A* turbojet engines with one engine mounted above and one below on the center line of the far aft end of the center section.

Crew:
• *Horten Ho 229 V6* - One person sitting inside a pressure cockpit cabin.
• *Gothaer Go P.60B* - Two people laying down (prone) side-by-side inside a pressure cockpit cabin.

Landing Gear:
• *Horten Ho 229 V6* - Tricycle with load distribution equal to 45% on the nose wheel and 55% on the two main wheels. Bulges in the upper surface of the center section streamline required.

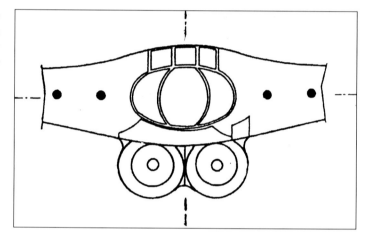

A pen and ink drawing featuring a head-on view of the repositioned *BMW 003* engines to center section's underside on the *Gothaer Go P.60A*. *Reimar Horten* told this author that he had considered a similar arrangement for their *Horten Ho 9*, however, he and his brother *Walter* were worried that the engines would suck up dirt from the runways and thereby significantly reducing engine life.

• *Gothaer Go P.60B* - Tricycle with load distribution equal to 15% on the nose wheel and 85% on the two main wheels. This was a customary fighter load distribution.

How the *Gothaer Go P.60A* with its twin *BMW 003* turbojet engines mounted on the under side aft of the center section might have looked in flight. Digital image by *Andreas Otte*.

With both of its BMW 003 turbojet engines mounted beneath its center section, the upper surface on the all-wing Gothaer Go P.60A would have been virtually absent of any speed robbing obstacles. Digital image by *Andreas Otte*.

Armament:
• *Horten Ho 229 V6* - *4x108 MK* (automatic machine cannon) with 90 rounds each, or *2x103 MK* (automatic cannon) with 140 rounds each.

• *Gothaer Go P. 60B* - *4x108 MK* (automatic machine cannon) with 170 rounds each, or *2x103 MK* (automatic cannon) with 175 rounds each.

Fuel Capacity:
• *Horten Ho 229 V6* - Four fuel tanks in each outer wing holding a total of 4,409 pounds (2,000 kilograms) of fuel.

• *Gothaer Go P.60B* - One fuel tank in the center section, plus one fuel tank in each outer wing for a total of 4,410 pounds (2,000 kilograms) of fuel. There is sufficient space within the center section for a 2nd fuel tank of 2,205 pounds (1,000 kilograms).

Aerial Photographic Equipment:
• *Horten Ho 229 V6* - 2 *Rb 50/15* photo reconnaissance cameras.
• *Gothaer Go P.60B* - 2 *Rb 50/15* photo reconnaissance cameras.

Construction:
• *Horten Ho 229 V6* - Center section: light metal tubular framework with plywood covering. Outer wing: wood spar, ribbed construction with reinforced wooden shell, and covered over with plywood.

• *Gothaer Go P.60B* - Center wing: metal tubular framework with plywood covering. Outer wing: wood spar, ribbed construction, lattice grate type construction with plywood covering.

Take-off Weight (with equal amounts of fuel):
• *Horten Ho 229 V6* - 18,739 pounds (8,500 kilograms).
• *Gothaer Go P.60B* - 16,520 pounds (7,450 kilograms).

How twin BMW 003s contained in a single nacelle might have appeared on the Gothaer Go P.60A when seen front-on.

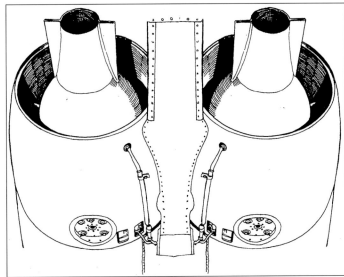

A pen and ink illustration of how twin *BMW 003s*, featuring their exhaust nozzles, might have appeared from behind and below on the all-wing *Gothaer Go P.60A*.

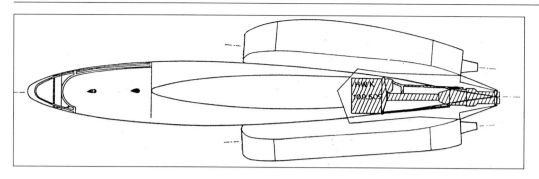

In another variation on the basic *Gothaer Go P.60A*, *Dr. Ruldolf Göthert* was considering the use of a single *Walter HWK 509B* bi-fuel liquid rocket engine to help the all-wing machine achieve a more rapid take-off. Installation of the *HWK 509B* would have been relatively simple: it would have been placed in the space between the upper and lower *BMW 003s*. This pen and ink drawing features the all-wing's port side.

Utilization of Other Propulsion Units:
• *Horten Ho 229 V6* - (*Jumo 004Bs* standard) installation of *BMW 003A* somewhat possible, but impossible for *Heinkel-Hirth HeS 011*.
• *Gothaer Go P.60B* - (*BMW 003A* standard) installation of *Jumo 004B* easily accomplished, as well as *Heinkel-Hirth HeS 011s*.

Wing Surface:
• *Horten Ho 229 V6* - 63.38 square yards (53 meters squared).
• *Gothaer Go P.60B* - 55.97 square yards (46.8 meters squared).

Overall Wing Surface:
• *Horten Ho 229 V6* - 142.32 square yards (119 meters squared).
• *Gothaer Go P.60B* - 131.56 square yards (110 meters squared).

Aspect Ratio:
• *Horten Ho 229 V6* - 5.2
• *Gothaer Go P.60B* - 3.3

Sweep Back:
• *Horten Ho 229 V6* - 28 degrees.
• *Gothaer Go P.60B* - 45 degrees.

Wing Sections:
• *Horten Ho 229 V6* - Asymmetrical standard wing sections. Ratio of maximum thickness: 17.8%, but originally 15.5 percent.
• *Gothaer Go P.60B* - Symmetrical high speed wing sections. Ratio of maximum thickness: 13.0 percent.

Elevator and Aileron:
• *Horten Ho 229 V6* - *Frise* control surfaces.
• *Gothaer Go P.60B* - Internally balanced control flaps.

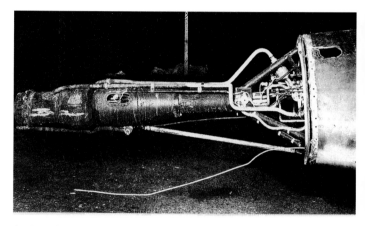

A photo featuring a single *Walter HWK 509B* experimentally installed in the tail of a *Messerschmitt Me 262* as seen from its starboard side. A similar arrangement would have been designed into the center section of the all-wing *Gothaer Go P.60A*...in the space between the over and under *BMW 003s*.

Directional Control:
• *Horten Ho 229 V6* - Interceptor-control in 80% of span (planned change in construction: control by resistance of surface sections being extended outward from the wing in direction of span; slot in the outer wing).
• *Gothaer Go P.60B* - On wing tip extendable plates perpendicular to the wing.

Elevator and Aileron Trimming:
• *Horten Ho 229 V6* - Inner control surfaces in cross position to outer control surfaces by kinematic superimposition (planned: auxiliary trim tab).
• *Gothaer Go P.60B* - Auxiliary trim tab in the main control surface.

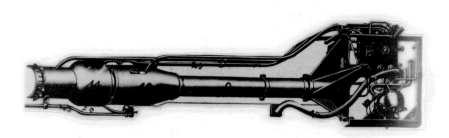

The starboard side of the typical *HWK 509B* bi-fuel liquid rocket engine from *Helmuth Walter*, Kiel.

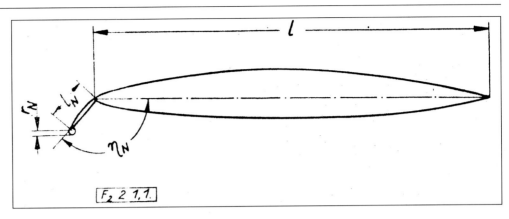

A pen and ink illustration from Gothaer Waggonfabrik featuring their plans for a laminar-flow wing profile to be used on the Gothaer Go P.60 series flying machines by *Rudolf Göthert*. Featured in this illustration is the innovative leading edge flap as seen to the left of the illustration.

Landing Devices:
• *Horten Ho 229 V6* - Simple trailing edge flap.
• *Gothaer Go P.60B* - A leading edge split flap with the split flap in wing mid-section.

Safety Devices Against Tipping:
• *Horten Ho 229 V6* - Slight warping of the outer wing.
• *Gothaer Go P.60B* - Slight warping of the outer wing, setting up wing sections along the span, and a leading edge split flap (alternative: automatic slot).

Position of Center-of-Gravity and Placement of Center-of-Gravity:
• *Horten Ho 229 V6* - Maximum permissible backward shift of the center-of-gravity at the take-off, attained only with a ballast of 1,322.40 pounds (600 kilograms). In reconstructing the wing area for series production, ballast is to be reduced considerably. In landing the center-of-gravity is shifted forward 7% farther than at take-off.
• *Gothaer Go P.60B* - Position of center-of-gravity is within the safety region.

Weight of Outer Wing With Aileron:
• *Horten Ho 229 V6* - 3,086.47 pounds (1,400 kilograms) and total area is 366 square feet (34 meters squared).
• *Gothaer Go P.60B* - 1,984.16 pounds (900 kilograms) and total area is 297.1 square feet (27 meters squared). This makes for a 19% smaller surface and 584.22 pounds (265 kilograms) less weight than the *Horten Ho 229 V6*.

Weight of Center Section:
• *Horten Ho 229 V6* - 2,447.13 pounds (1,110 kilograms) and 205 square feet (19 meters squared).
• *Gothaer Go P.60B* - 2,017.23 pounds (915 kilograms) and 213 square feet (19.8 meters squared). These differences provide for a considerably improved static construction, the avoidance of eccentricities, and 429.90 pounds (195 kilograms) less weight than the *Horten Ho 229 V6*.

The port outer wing of a Horten Ho 7 featuring its experimental extendable drag rudder. This type of drag rudder was to be tested on the *Horten Ho 7*, however, it was never tried out in flight. *Dr. Rudolf Göthert* was considering a similar drag rudder arrangement on his *Gothaer Go P.60* flying machines.

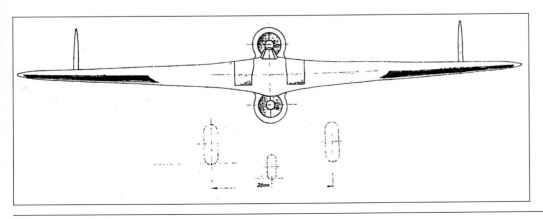

A pen and ink illustration featuring the *Gothaer Go P.60B* tailless flying machine. The darkened areas on the wing's leading edge are *Göthert's* innovative "leading edge flaps."

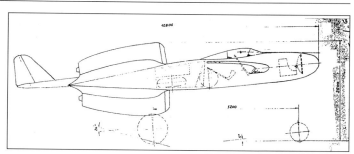

A poor quality pen and ink starboard side view of the *Gothaer Go P.60C* featuring positions of its three man crew and with its internal nose mounted *FuG 240* centimetric scanning radar.

The experimental extendable wing tip drag rudder seen here on a *Horten Ho 7* fully retracted.

Weight of Twin Turbojet Engines:
• *Horten Ho 229 V6* - (*Jumo 004B*) 3,825.02 pounds (1,735 kilograms).
• *Gothaer Go P.60B* - (*BMW 003A*) 3,461.26 pounds (1,570 kilograms). Approximately 363.76 pounds (165 kilograms) less weight than the *Horten Ho 229 V6*.

Weight of Ballast:
• *Horten Ho 229 V6* - 1,322.77 pounds (1,735 kilograms).
• *Gothaer Go P.60B* - Zero ballast and providing 1,322.77 pounds (600 kilograms) less weight than the *Horten Ho 229 V6*.

Weight of Armor Plating:
• *Horten Ho 229 V6* - 881.85 pounds (400 kilograms).
• *Gothaer Go P.60B* - 639.34 pounds (290 kilograms) for 242.51 pounds (110 kilograms) less weight.

Weight of Landing Gear:
• *Horten Ho 229 V6* - 760.59 pounds (345 kilograms).
• *Gothaer Go P.60B* - 1,157.43 pounds (525 kilograms) and pro-

viding for 396.83 pounds (180 kilograms) more weight than the *Horten Ho 229 V6*. Landing gear is from a *Junkers Ju 88*.

Weight of Ammunition:
• *Horten Ho 229 V6* - 462.97 pounds (210 kilograms).
• *Gothaer Go P.60B* - 859.80 pounds (390 kilograms) and providing 396.83 (180 kilograms) more weight than the *Horten Ho 229 V6*.

Weight of Pressure Cockpit Cabin:
• *Horten Ho 229 V6* - No pressure cabin provided.
• *Gothaer Go P.60B* - 264.55 pounds (120 kilograms) and providing 264.55 pounds (120 kilograms) more weight than the *Horten Ho 229 V6*.

Overall Weight Difference Between The *Horten Ho 229 V6* and the *Gothaer Go P.60B*:
• The *Gothaer Go P.60B* would have weighed 2,403.04 pounds (1,090 kilograms) less than the *Horten Ho 229 V6*.

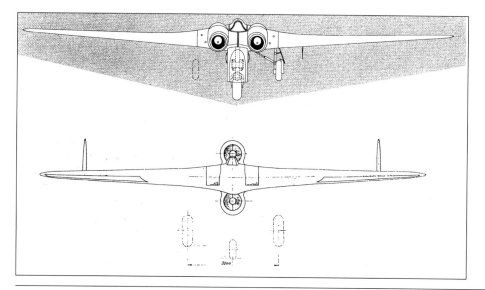

A nose-on comparison of the all-wing *Horten Ho 229 V3* and the tailless *Gothaer Go P.60B/C*. These pen and ink illustrations are not to scale. Does it appear that the *Horten Ho 229 V3* had the cleaner profile?

Review of Anticipated Flight Performances and Flight Characteristics Between the *Horten Ho 229 V6* and the *Gothaer Go P.60B*

By *Gothaer Waggonfabrik AG**

In this present elaboration (January 27th, 1945) we shall try to compare the two airplanes: the *Horten Ho 229 V6* and *Gothaer Go P.60B*, which is an advanced model of the former, in their most important structural and aerodynamical basic layout.

For comparison, a model of the *Horten Ho 229 V6* has been used, since the latter contains all of the required installations. It is to be mentioned specifically that because of this, the *Horten Ho 229 V6* will still seem less favorable by this comparison; compared to the original design specifications (*Horten Ho 9 V2*) changes were necessary in the all-wing *Horten Ho 229 V6* toward a change for the worse in aerodynamics and weight. (It should be noted, too, that *Gothaer Waggonfabrik* was going to install wing tip vertical surfaces with an attached hinged rudder...similar to their proposed *Gothaer Go P.60C*, and thus, in effect, making their *Gothaer Go P.60B* a tailless flying machine.)

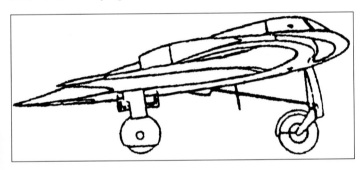

A pen and ink drawing of the starboard side of the *Horten Ho 229 V3*. Not to scale. Unfortunately, no known wind tunnel comparisons exist between the *Gothaer Go P.60C* and the *Horten Ho 229 V3* by which drag coefficients could be determined.

Both aircraft, the *Horten Ho 229 V6* and the *Gothaer Go P.60B*, are to be considered as heavy fighters, single seat or two-man, respectively. They would be required to carry out reconnaissance for both close and intermediary distances.

The fact that the two compared airplanes are equipped with different propulsion units must be emphasized before a discussion of fight performance can take place. The difference in power (thrust) is of practical importance only at ground level or at low altitudes, and are especially of disadvantage to the *Gothaer Go P.60B*, since they reduce its power by 6.6 feet per second (2.0 meters per second) on the ground and on the runway by 393.7 feet per second (120 meters per second).

By transferring the propulsion units to the outside of the machine and by adopting an arrangement where the pilot is in a lying (prone) position, it becomes possible to keep the wing area in the *Gothaer Go P.60B*, in contrast to the *Horten Ho 229 V6*, about 12% smaller, and to reduce the maximum thickness of the wing section to 13% versus 17.8 percent. Beyond that, symmetrical wing sections with a small leading-edge radius and a shift of the bulk of the flying machine toward the stern (50%) are being used in the *Gothaer Go P.60B*; the sweep back is also considerable. Because of these measures (small thickness of wing sections, small leading-edge radius, considerable shifting of thickness to the rear, strong sweep back, and also small aspect ratios), it is expected that the start of the sharp rise in drag due to the approach of the flight speed to sonic velocity will be shifted in the *Gothaer Go P.60B*, as against the *Horten Ho 229 V6*, to larger *Mach* numbers.

*Comparisons Of the *Horten 8-229* and the *Go P-60* All-Wing Airplanes

The *Horten Ho 229 V3* was steered by two drag rudders extending out (above and beneath) near each wing tip. On the *Horten Ho 229 V3*, these drag rudders were incremental, that is, the outer drag rudder extended first, then followed by the inner drag rudder. When a tighter turn was required then the pilot could extend the inner, too, for maximum effectiveness.

The *Gothaer Go P.60*, being tailless, did not require the use of drag rudders. Featured in this photo, the wing tip drag rudders on the *Horten Ho 229 V3* are fully open...above and beneath the wing...for maximum effectiveness. *Dr. Rudolf Göthert* believed that the use of drag rudders on the high speed all-wing would not correct its directional instability and that a conventional rudder like on his proposed *Gothaer Go P.60B/C* was absolutely necessary.

The larger induced drag of the *Gothaer Go P.60B*, effected by the lower aspect ratio, caused slight losses in climbing speed and ceiling which, however, is compensated by the smaller loaded weight.

The additional drag, produced in a *Gothaer Go P.60B* by removing the propulsion units from the wing, is partly balanced by the smaller wing area and by using thinner symmetrical wing sections. Beyond that, the magnitude of this additional drag depends largely on the possibility of transferring the propulsion units into the wing. The wing sections selected for the *Gothaer Go P.60B* are decidedly laminar sections. Should these wing sections still show laminar effects at the large *Reynolds* numbers with which we deal here, then the attainable *cw F*-value (without *M*-influence) in the region of high speed will be smaller than in the *Horten Ho 229 V6*. However, since the *Horten Ho 229 V6* as well as the *Gothaer Go P.60B* perform at speeds where sound velocity is approached and where, consequently, a sharp rise in drag must be considered, this drag will decisively influence the maximum flight performances. Consequently, the *Gothaer Go P.60B* is expected to show definite advantages over the *Horten Ho 229 V6*.

Flight performances of both airplanes, as plotted in *Figure N*, were ascertained under the same basic assumptions (*DVL*-Berlin measurements of wing sections, equal *cw* - additions for landing gear flaps, control surface slot, and good construction of surfaces):

the increase of drag at high *Mach* numbers was considered according to the high speed measurements by *DVL*. Shift of the critical *M*-number by sweep back in the *Horten Ho 229 V6* was ÉM=0.03, in the *Gothaer Go P.60B*, ÉM=0.05.

Flight Characteristics

Since, in high speed airplanes investigated thus far, the dangerous disturbances of stability around the lateral axis start only at *M*-numbers higher than those at which the sharp rise in drag sets in, it is to be expected that, at least in horizontal flight and climb, the disturbances in flight characteristics around the lateral axis will also follow the same tendency in tailless aircraft, especially since the outer control surfaces might become "critical" much later because of their thinner air foil sections.

In the *Gothaer Go P.60B*, compared with the *Horten Ho 229 V6*, the symmetry, practically 100%, of suction side and pressure side might show up favorably, since any asymmetry of wing sections or of the overall structure of an airplane produced stability disturbances around the lateral axis in comparatively early stages of high speed performance.

Since in a tailless aircraft the direction stability, and especially the stability in turning and banking is rather weak, attention will have to be focused on this problem in discussing these models. In the *Gothaer Go P.60B*, improvements are expected by placing the

A nice view of the drag rudders (upper and lower) on the *Horten Ho 229 V3*'s starboard wing.

A pen and ink illustration of *Rudolf Göthert's Gothaer Go P.60C* night fighter featuring its nose port side. This tailless machine would have featured the *FuG 240* internal nose mounted centimetric wavelength airborne interception radar, and would replace the all-wing *Horten Ho 229 V6* night fighter.

A close up view of the starboard wing tip drag rudders on the Horten Ho 229 V3, which in Rudolf Göthert's opinion were insufficient for the pilot to maintain directional control over the large all-wing.

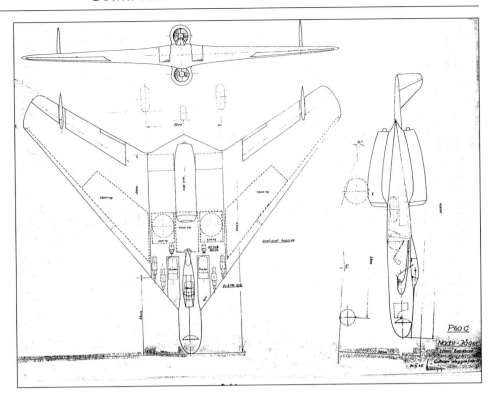

A pen and ink three view drawing featuring the three man *Gothaer Go P.60C* night fighter project.

propulsion units backward and outward. Special measures have been planned in both flying machines for target approach. In order to increase the stabilizing effect while making a turn the *Horten Ho 229 V6* extended its air-wing brakes, while the *Gothaer Go P.60B* used its wing tip plates, which also were to be used for directional control. However, installation of an automatic target flight control was recommended for both flying machines. Nevertheless, adequate directional stability in the *Gothaer Go P.60B* would have been attained by merely installing vertical tail surfaces with an attached rudder similar to that proposed for the *Gothaer Go P.60C*.

Gothaer Waggonfabrik believed vertical control surfaces and, consequently, the tactical brake flaps of the *Horten Ho 229 V6* would be liable to present difficulties at high speed as a result of upsetting the entire distribution of pressure along a considerable part of the wing span. In this respect, it may be more favorable to adopt brake surfaces extendable at the wing tip, although the wide gap, forming at the wing tip, was thought to be a disadvantage.

An accurate coordination of the controlling components was exceedingly difficult due to the large dimensions of the control surfaces and the high speed of flight, for the required aerodynamic balance of control surface moments must be carried so far that the unavoidable allowances, to be made in production, will make the control, without special measures, practically impossible. In the *Horten Ho 229 V6* special difficulties might be expected in that respect, chiefly at high velocities, because extreme *Frise*-type control surfaces were adopted.

In the *Gothaer Go P.60B* these difficulties would be alleviated by subdividing control surfaces, whereby only the smaller outer

The tailless *Gothaer Go P.60B* tandem seat day fighter featuring its starboard side. Beneath its wings hang four *Fritz "X"* air to air wire-guided missiles. *Göthert* believed that its wing tip vertical stabilizers with attached rudders would prove it to be superior to the drag-rudder steered *Horten Ho 229 V3* and its well-known directional control problems (side to side oscillations).

Dr. Rudolf Göthert would have placed twin *Heinkel-Hirth HeS 011A* turbojet engines in his *Gothaer Go P.60B*. He calculated that his tailless machine would have had a top speed of 609 miles per hour [980 km/h] verses 587 miles per hour [945 km/h] for the twin *Jumo 004B* powered *Horten Ho 229 V3*.

control surface is directly controlled by the stick, while the main control surface would be indirectly controlled by the stick and moved by a servo-rudder.

Because of the heavy sweep back, stability was believed to be assured more difficult to achieve in a *Gothaer Go P.60B* than in a *Horten Ho 229 V6*. But here, too, satisfactory performance characteristics were to be expected by applying the usual remedies against tipping hazards, such as: the traditional methods of fairing along the span, leading edge split flap, or slot.

The landing of tailless aircraft naturally calls for placing the center-of-gravity as far to the rear as possible. In the *Gothaer Go P.60B* this will be done by arranging the landing gear, fuel tanks, and ammunition to that effect; the center-of-gravity is slightly shifted to the rear when the landing gear is lowered, the fuel tanks become empty, and the ammunition is expended. In the *Horten Ho 229 V6* conditions are less favorable, since fuel, ammunition, and landing gear are disadvantageously arranged in regard to the position of the center-of-gravity at landing. Compared to the take-off, the situation on landing shows the position of the center-of-gravity to be further in front (approximately 7% of the reference chord = 200 mm, or 7.784 inches). The *ca max* values are still attainable under these adverse circumstances, but are very small and, consequently, the landing speed increases considerably.

Above: A close up view of the starboard side of the internal nose mounted *FuG 240* centimetric interception radar.

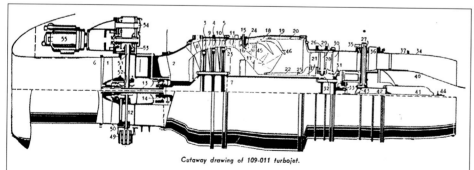

Cutaway drawing of 109-011 turbojet.

1. Front axial rotor	21. Guide grip in front of turbine wheel	38. (Not shown—not identified)
2. Diagonal rotor	22. Circular chamber	40. Thrust nozzle sleeve
3, 4, 5. Third, fourth, fifth rotors	23. Annular cool air gap	41. Piston rod
6. Air intake hood	24. Injection nozzles	42. Pressure cylinder
7. Main shaft	25. Cool air passage to guide grids	43. Piston
8, 9, 10. Guide grids	26. Fastening flange	44. Piston rod bearing
11. Triple grid	27. Connection for lubricant pipe	45. Mixers
12. Pump drive shaft	28. Guide grids between turbine rotors	46. Scoops
13. Floating rod	29, 30. Turbine rotors	47, 48. (Not shown—not identified)
14. Front main bearing	31. Cool air holes	49. Lubricant pump set
15. Connection for fuel pipe	32. Cool air holes	50. Front compressor housing
16. Arrow-shaped ring (nozzle ring)	33. Rear main bearing	51. Vertical shaft
17, 18. Air flows between compressor and turbine	34. Thrust nozzle housing	52, 53. Bevel gears
	35. Cool air connection	54. Drive connection for turbine's extension shaft
19. Housing wall	36. Ventilation tube	55. Starter
20. Circular chamber	37. Pressure measurement place	

A pen and ink illustration of the 2,866 pound [1,300 kilogram] thrust *Heinkel-Hirth HeS 011A*, as well as a description of its component parts. The *HeS 011A* was designed to produce 2,640 pounds thrust initially, which would ultimately increase to 3,250 pounds thrust at 560 miles per hour at sea level.

Dr. Reimar Horten Remembers *Dr.-Ing*. *Rudolf Göthert*
As Told to *Dr. David Myhra*

At *Gothaer Waggonfabrik AG* there was an aerodynamicist named *Dr.-Ing. Rudolf Göthert*. He had come to *Gothaer* from *LFA-Braunschweig* with considerable experience with the wind tunnels located there, and before that from *DVL.Berlin-Adlershof*. But he was not a pilot...only experienced at placing scale model aircraft in highly turbulent wind tunnels. *DVL* was supported financially by the *RLM*. *Göthert* had been conducting wind tunnel experiments on the aerodynamics of swept back wings and control surfaces for use in high speed flight. His real speciality was control surfaces, such as ailerons, flaps, rudders, and elevators. For example, *Göthert* would place a scale model with a swept-back wing in *DVL's* wind

tunnel and then measure the aerodynamics of the model with the swept back wings. But these measurements just could not be correct, for the moments were different during the angle of attack. When my brother *Walter* and I learned of his method of testing and then read the results of his wind tunnel investigation, well, we both had a good laugh. *Göthert* was completely wrong.

Göthert transferred to *Gothaer* from *LFA*. I don't know how or why this transfer came about. But afterward he claimed our all-wing *Horten Ho 229 V6* would experience directional stability problems. In our *Horten Ho 9 V1* and *Horten Ho 9 V2*, as well as our other all-wing sailplanes, we learned that the lift distribution was in

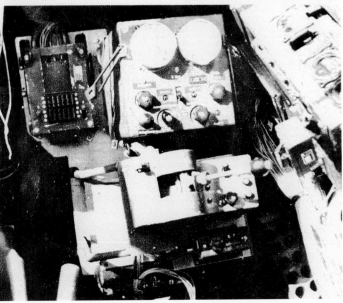

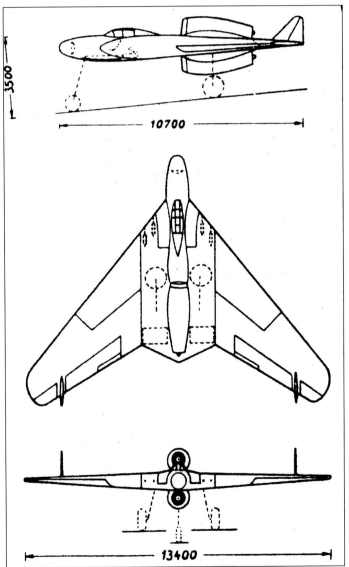

Upper photo: a port side view of the *FuG 240* centimetric interception radar comparing its overall size to the RAF individual to the right in the photo. Lower photo: radar operator's *FuG 240* instrumentation. This is similar to what the *Rudolf Göthert*-designed *Gothaer Go P.60B* would have included in its tandem cockpit.

A pen and ink three view illustration of *Rudolf Göthert's* tailless *Gothaer Go P.60B* three man day fighter. It had a wingspan of 44 feet 3 inches verses 40 feet 6 inches for the all-wing *Gothaer Go P.60A*.

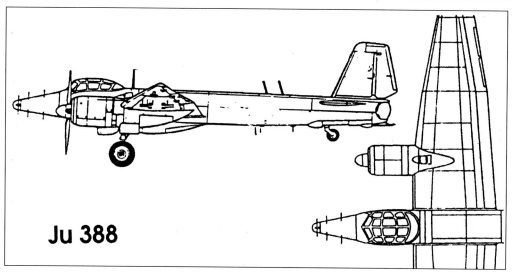

Ju 388

A pen and ink drawing of the Junkers *Ju 388J3* twin engine bomber featuring the *FuG 240* internal nose mounted interception radar.

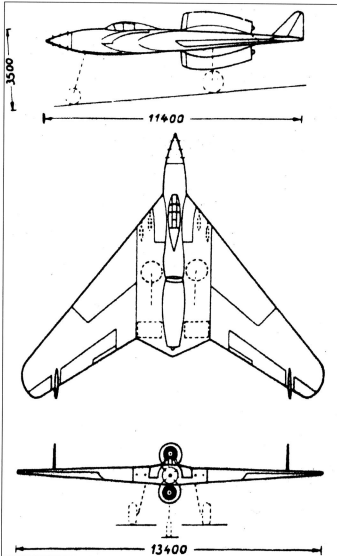

A three view pen and ink illustration of the *Gothaer Go P.60C's* planform. This version would have been largely identical to the *Gothaer Go P.60B* with the exception of its fuselage nose. It would have been more pointed since the *FuG 240* interception radar was not included.

the form of a bell curve. That is to say, that the wing needed a twist in its tapering. This twist was 1.0 *CL*, and for our *Horten Ho 9 V1* the twist was 0.3 *CL*. Later on we obtained about 0.6 *CL*. This twist had been ordinarily achieved by the aileron. I had determined that the center of lift would be achieved in the same place that corresponded to the flying machine's center of gravity. We called the condition the "*Polar of Equilibrium*." I discovered that this equilibrium would occur at all *CL* until one reached the maximum.

Now, *Berhard Göthert* had measured a swept back wing in his *DVL* wind tunnel without ailerons and gave the flying machine a *CL* maximum. This could not be. The air flow would be completely different in the way *Berhard Göthert* measured the aerodynamic effects of the scale model in his *DVL* wind tunnel. Consequently, we *Hortens* did not take him or his *DVL* work seriously. Then there was another thing. *Rudolf Göthert* was investigating flaps. He had placed flaps along the entire trailing edge of the wing. We had, in the outer wing, flaps working up but not down. Therefore, the moments created by *Berhard Göthert* at *DVL* were completely different than what we had experienced. So, the moments *Berhard Göthert* experienced in the *DVL* wind tunnel were not good. In addition, the twists were reversed. The result of all the *Göthert* brother's experiments with the *CL* maximum resulted in a total change of the wing tips.

There were other bad things which *Berhard Göthert* did with his scale model in the *DVL* wind tunnels. For example, if *Rudolf Göthert* had only placed the flap in the center section, then he might have gotten results similar to our experiences...perhaps even better. Perhaps the moment would have been zero. So *Berhard Göthert* was testing a swept back wing in his *DVL* wind tunnels, but he had no practical application, and consequently his test results were without use to us or anyone else. All this was happening about August 1944.

When all *Berhard Göthert's* information was being sent out from *DVL*, my brother *Walter* and I felt that it was a great pity that we could not have spoken to this man earlier and told him what glaring errors he had made. However, *DVL*, as well as *Dr. Rudolf Göthert*, felt that they were in such a position that they would not accept any criticism, especially from people like us...people who

An overhead view of a *Gothaer Go P.60C* with a projectile nose and featuring its port side. The image in this photo is a scale model.

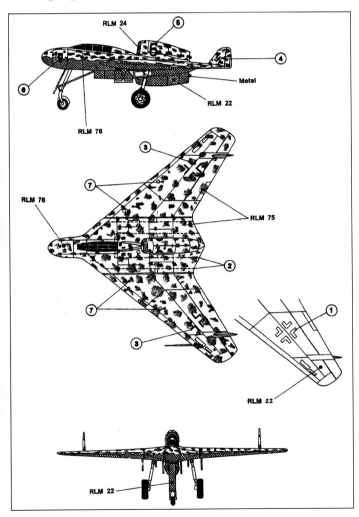

A pen and ink drawing of the *Gothaer Go P.60C* featuring its proposed camouflage:
- *RLM 22* - black   - *RLM 75* - gray-violet
- *RLM 24* - red   - *RLM 76* - light blue

had no formal training. Most of the *Göthert* twins' work was completely out of date, because his method of placing scale models in highly turbulent wind streams and so on had been shown to give false readings.

I saw different test results, too, with *Alexander Lippisch's* delta wing designs. These wind tunnel results, like those of *Rudolf* and *Berhard Göthert*, were false, too. Whenever we read that results of *Berhard Göthert's DVL* wind tunnel results, our only response was laughter. It was all a big joke, because all these men who ran the wind tunnels had no practical experience like *Willy Messerschmitt, Alaxander Lippisch*, or the *Horten* brothers. The stuff which the *Göthert* brothers ran in the *DVL* wind tunnel did show good results, but were practically worthless if we tried to apply them on a practical basis. So the money and time spent by *DVL's Berhard Göthert* and *LFA's Rudolf Göthert*, and at other wind tunnels in *Nazi* Germany was a waste. After *Walter* and I read the summary of the

The asymmetrical *Blohm und Voss Bv 141. Rudolf Göthert* told *AAAFI* agents that his proposed *Gothaer Go P.60* series would have utilized an electrical aileron (elevon) device to produce artificial lateral stability as used on the asymmetrical *Bv 141. Blohm und Voss* achieved this by placing two small vanes, both with different moment arms, mounted on the lower surface of the wing...one pair ahead of each aileron.

A three view presentation of the proposed all-wing *Gothaer Go P.60A* day fighter by *Dr. Rudolf Göthert*. It was to have been powered by twin *BMW 003A* turbojet engines mounted above and below the aft center section. In addition, *Göthert* claimed that a single *Walter HWK 509B* bi-fuel liquid rocket engine could be installed in the space between the over and under *BMW 003A* turbojets. Digital image by *Andreas Otte*.

Göthert brother's work we simply put it aside and never went back to it again. Then we learned that *Rudolf Göthert* had transferred to *Gothaer Waggonfabrik* about August 1942.

We *Horten* brothers had no influence with the people who ran the wind tunnels in Germany. Although we were friends with *Professor Ludwig Prandtl* at Göttingen, he had no authority in scheduling his wind tunnel. The people who had the power included:

• *Dr. Dötsch* - DVL - profiles
• *Professor Quick* - DVL - general configuration and propellers
• *Dipl.-Ing. Liebe* - DVL - wing tip stall
• *Dipl.-Ing. Hohler* - DVL -spinning

• *Professor Seifurt* - AVA - high lift devices
• *Dr. Kuchemann* - AVA - special engines
• *Dr. Bernard Göthert* - DVL - high speed problems
• *Professor Schlichting* - *Göttingen* - interference
• *Dr. Rudolf Göthert* - *Gothaer Waggonfabrik AG* - tail surface controls
• *Dr. Liebe* - ailerons
• *Dipl.-Ing. Roos* - flutter and vibration
• *Obering Gotthold Mathias* - *Focke-Wulf* - directional and lateral stability
• *Dipl.-Ing. Hans Multhopp* - *Focke-Wulf* - longitudinal stability
• *Dr. Rudolph Schmitt* - *Dornier* - air brakes

Rudolf Göthert designed this hoped for replacement to the *Horten Ho 229* to be flown in a prone position, thus it presented usually clean lines. Its wing sweep back was **45 degrees** at its quarter chord. Digital image by *Andreas Otte*.

Although an all-wing design, *Rudolf Göthert* fixed a pair of "stabilizing fins" above and below each wing tip for directional control. This *Gothaer Go P.60A*, in gray camouflage, is seen from its port side. Scale model and photographed by *Reinhard Roeser*.

- *Dr.-Ing.* - *Schmitz* - *Heinkel AG* - air loads
- *Dipl.-Ing.* - *Puffer* - *Messerschmitt AG* - swept wings
- *Dr. Hoener* - *Messerschmitt AG* - drag
- *Professor Weise* - *Messerschmitt AG* - engines and cooling

Board members of the powerful *Sonderausschusses Windkanale* included:

- *Professor Quick*
- *Professor Seifurt*
- *Dipl.-Ing. Eicke*
- *Dr. Backhaus*
- *Dr. Conradis*
- *Dipl.-Ing. Hueber*
- *Dr. Motzfeld*
- *Dipl.-Ing. Schomerus*

The proposed *Göthert* designed all-wing day fighter *Gothaer Go P.60A* as seen from above. It featured "elevons," that is, combined elevators and ailerons. Scale model and photographed by *Reinhard Roeser*.

Reimar Horten Recalls the Göthert Twins and Gothaer Waggonfabrik

The name *Gothaer Waggonfabrik AG* as a possible manufacturers of our *Horten Ho 229* series came up at the *RLM* in the Summer of 1944. The *RLM* was thinking of series production of our flying machine. They were waiting for the first flight tests, which had started in December 1944 at Oranienburg. It was at this time *Gothaer Waggonfabrik* received an order to construct the pre-series *Horten Ho 229 V3, V4, V5,* and *V6*. But *Gothaer* had not done any work for us before they received the *Horten Ho 229* pre-series order from the *RLM*. As I mentioned, *Gothaer* was not mentioned to us until about August 1944. *Walter* mentioned to me that he was thinking about using *Gothaer* to manufacture the *Horten Ho 229* series. I told *Walter* that it did not matter to me which aviation company was picked to build the *Horten Ho 229*, because I was not interested in that part of aircraft production. My main interest was in

The *Gothaer Go P.60A* would have had a wing area of 504 square feet [46.8 square meters] and special leading edge flaps to improve stall characteristics. Water color by *Loretta Duval*.

The all-wing *Gothaer Go P.60A* as seen from its starboard side. The prone piloting position gave the flying machine very nice lines, however, it would have presented serious difficulties due to its restricted rearward field of vision. Digital image by *Andreas Otte*.

The all-wing two man *Gothaer Go P.60A* was designed with one degree of dihedral and 45 degrees of sweep back at its quarter chord. Digital image by *Andreas Otte*.

design and development work. Once the project was about to go into production I did not want to be involved any longer, because I had other projects which required design and development. It was now the *RLM's* responsibility to see how and where our *Horten Ho 229* series production would take place and by whom. I guess that *Gothaer* had had some practical experience in wood construction so they had been selected by *Walter* and the *RLM*.

Klemm Flugzeugbau had also been selected as a constructor of what was designated the *Horten Ho 229*. We knew *Fritz Klemm* personally. He had built our *Horten Ho 7*, plus he had had considerable experience in building light planes, and *Walter* and the *RLM* had selected *Klemm* at the same time they picked *Gothaer*. The

thing with *Gothaer Waggonfabrik* is that during the war they had built for *DFS* their *DFS 230* troop transport glider. But *Gothaer* had changed *DFS'* design...not aerodynamically, but in its structure before they started series production. These modifications made by *Gothaer* were not requested by *DFS*, the troops, or anyone else. *Gothaer* made structural modifications in order to ease construction...to make construction easier and faster. As a result, several *DFS 230* troop transport gliders had broken up in flight. So I told *Walter*, "I want you to observe *Gothaer* during the construction of our pre-production *Horten Ho 229 V3* through the *V6* so that they don't make any modifications which might harm our aircraft during its flight testing, just for the sake of more rapid produc-

A pair of two man all-wing *Gothaer Go P.60A* day fighters. Armament would have included 4x*MK 108 30mm* cannon mounted in the center section...two on each side of the cockpit. Digital image by *Andreas Otte*.

Two *Gothaer Go P.60A* all-wing day fighters. *Rudolf Göthert* stated post war that his day fighter had a proposed range of 995 miles [1,600 kilometers] verses 870 miles [1,400 kilometers] for the *Horten Ho 229*. Digital image by *Andreas Otte*.

The camouflage of this all-wing Gothaer Go P.60A is a combination of RLM 75 and RLM 76. Digital image by Andreas Otte.

At the time *Gothaer* received the order to construct four pre-production *Horten Ho 229s* by the *RLM*, they had very little free construction space. So they squeezed the *Horten Ho 229* into a small work space at Friedrichroda, where twenty people were working on these four pre-production machines. *Mr. Büna* was there at Friedrichroda watching what the *Gothaer* people were thinking about modifying for series production. The chief of *Gothaer's* small group was a man whom I knew. He was *Mr. Kaufmann*, and was a very good man, too. I could easily talk with him, and he assured me that he had constructed these four *Horten Ho 229* pre-production machines just the way we had designed them. So with *Büna* there at Friedrichroda as our secret inspector and my friendship with *Gothaer's* construction chief *Kaufmann* on the *Horten Ho 229* project, I had no reason, therefore, to expect difficulties from *Gothaer*.

tion." If they did, just for the sake of faster construction, and one or more later crashed, well, it would be bad for the *Horten* brothers as the designers of all-wing aircraft. So *Walter* sent a graduate engineer by the name of *Büna* to *Gothaer* to watch them. *Gothaer* would have to get permission from *Büna* before they could make any change or modification merely for the sake of more rapid construction. This way *Gothaer* could not change structural items for series production which might later weaken the *Horten Ho 229* in flight to the point where adverse criticism might be heaped on the *Horten* brothers and our all-wing designs.

Getting back to *Dr. Rudolf Göthert*, well, at *Gothaer* they though that they could hire any man to carry out aircraft design. *Rudolf Göthert* had gone to *Gothaer Waggonfabrik* without any prior practice of glider building or flying, and without any experience in designing motored aircraft. His whole experience in aviation had been an academic one. His whole experience was investigating different aircraft shapes in wind tunnels. We *Horten* brothers knew that this academic experience would not come to a good end.

We later learned that *Rudolf Göthert* knew nothing of our *Horten Ho 9's* design history, in fact, nothing of all our all-wing experiments going back ten years to our *Horten Ho 1*. So he knew nothing of all our experimentations leading to the *Horten Ho 229*. He knew nothing about my work in wing lift distribution. See, that was the thing! *Rudolf Göthert* had only seen our *Horten Ho 9 V1* and the *Horten Ho 9 V2*, and then he now came out with a similar

Late in the war *Luftwaffe* heavy day fighters, like this proposed *Gothaer Go P.60A*, would be camouflaged in great variety. Digital image by *Andreas Otte*.

In addition to the twin *BMW 003As* of 1,763 pounds [800 kilograms] thrust turbojet engines called for in the design specifications, a single 4,400 pound [1,9165 kilograms] static thrust *Walter HWK 509B* bi-fuel liquid rocket engine could also be installed to give the all-wing superior climbing ability. Digital image by *Andreas Otte*.

The two crewmen in the *Gothaer Go P.60A*—pilot and radio/navigator—flew the all-wing in a prone position in a pressurized cockpit cabin. Digital image by *Andreas Otte*.

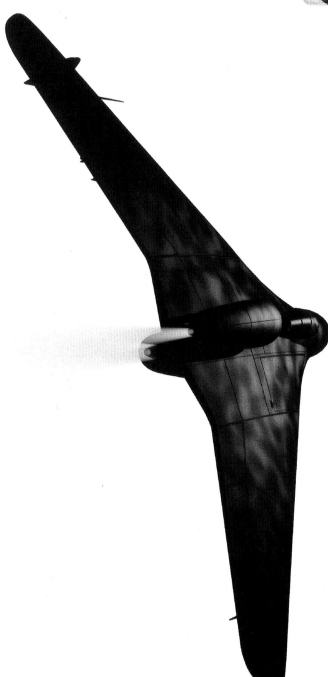

design at *Gothaer* which he was calling the *Gothaer Go P.60B*. I do not believe that *Rudolf Göthert* knew why he was designing *Gothaer* all-wing projects the way he did. Really, the *Gothaer Go P.60B* would have been a tailless machine. Look at the photographs of this proposed flying machine. The wings have a certain tapering which he copied from our *Horten Ho 9 V2*. He made the wing root on his *Gothaer Go P.60B* greater than on our *Horten Ho 9 V2*...I suppose to make it more easy to construct. Overall, his *Gothaer Go P.60B* is very similar to that of our *Horten Ho 229*, but with those vertical stabilizers/rudders I seriously doubt that the *Göthert* twins understood the aerodynamics behind our design and how our *Horten Ho 229* came to be.

I never spoke to *Rudolf Göthert*, nor his twin brother *Berhard*. I had seen *Rudolf Göthert* from time to time at different functions and conferences that we both attended, but I never sat down with him and talked about aircraft design...especially with our *Horten Ho 9 V2*. *Rudolf Göthert* did not appear to be interested in how we *Horten* brothers had designed and built our *Horten Ho 9 V2*...and I was not about to teach him. You see, we *Hortens* had come from different backgrounds than this man. However, if a man such as *Rudolf Göthert* would have come to me and asked me "how did you come to design your all-wing *Horten Ho 9*?" then I would have welcomed him and would have clearly discussed the *Horten Ho 9 V2* design history with him. But if a person showed complete disinterest like *Rudolf Göthert* did, then I would not brother to approach him. And why should I have? Also, I did not believe that the *Göthert* twins could have designed the so-called *Gothaer Go P.60B* all by themselves, because neither one had any prior design experience that I am aware of. Instead, *Rudolf Göthert* copied our *Horten Ho 9 V2*. What we see in the *Gothaer Go P.60B* is a picture of only a proposed flying machine. A paper design. Any school boy can draw a picture of an aircraft. But to build one *Rudolf Göthert* needed experience and a design staff to make it as it should be. All this stuff must be learned through trial and error like the old timers, such as *Willy Messerschmitt, Ernst Heinkel, Alexander Lippisch*, and later we *Horten* brothers.

How do I think *Rudolf Göthert's Go P. 60B* would have performed...you are probably wondering? Well, my opinion is that if it had been built, it would have performed about the same as our

The proposed *Gothaer Go P.60A* seen in a banking turn at full power. Maximum level speed was calculated to be 568 miles per hour [915 km/h] at 20,000 feet altitude. Digital image by *Andreas Otte*.

A ground level view of the proposed heavy day fighter *Gothaer Go P.60A* as seen from its nose port side. All fuel would have been carried in the outer wings. Digital image by *Andreas Otte*.

Horten Ho 229. Since *Rudolf Göthert* had copied the profile of our *Horten Ho 9 V2*, the *Gothaer Go P.60B*'s performance would have also depended on the thrust of the turbojet engines. He wanted to use *BMW 003s*, while we were using *Jumo 004s*. Then again, *Rudolf Göthert* wished to place one *BMW 003* on top and one on the bottom of the aft center section. This could not be. This arrangement we had for a time with one of our *Horten Ho 9* concept ideas. We reasoned that if a turbojet engine would be mounted on the under side of the center section it would pick up sand and dirt from the runway, and its operational life would be shortened considerably. Then again, placing the turbojets on the outside along the top of the center section is an idea which we had for our *Horten Ho 10A* and *Ho 10B* delta all-wing fighters. I believed that this arrangement was less desirable than putting the turbojet engines inside the wing. See, we had thought through all these different engine locations.

Rudolf Göthert also placed so-called stabilizing fins near the wing tips on his proposed *Gothaer Go P.60A* and vertical stabilizers with attached rudders on his tailless *Gothaer Go P.60B* and *Go*

Dr. Rudolf Göthert proposed a variation to his *Gothaer Go P.60A*. In this variation the all-wing's upper and under *BMW 003A* turbojet engines were changed so that both turbojet engines were under the center section. Digital image by *Andreas Otte*.

Rudder control on the Gothaer Go P.60A was to have been achieved by telescoping a surface above and below the wing tip, thereby creating drag. These telescoping surfaces were similar to those in a *Horten Ho 7*, however, on the *Ho 7*, the telescoping surfaces extended horizontally out of the wing tip. Digital image by *Andreas Otte*.

The upper surface of the *Gothaer Go P.60A*, with both of its *BMW 003As* now located under the center section, gave this all-wing a very aerodynamically smooth top surface. Digital image by *Andreas Otte*.

P.60B designs, as we see in these drawings. But when he was ready to construct a prototype it would have carried vertical stabilizers/rudders like those found on his proposed *Gothaer Go P.60C*. I guess he initially reasoned that this was a good place to provide drag. Really, he could have used any form of surface. We could have done it too, on our *Horten Ho 9 V2*, but we chose to keep this half of the wing free of any obstacles. We chose instead to have our drag devices located inside the wing, and then when it was needed

it would appear out of the wing to produce drag. *Rudolf Göthert's* fins would give a high degree of drag all the time the machine flew...whether it was needed or not. I believe that his choice of fins near the wing tip was not as good as our system on the *Horten Ho 9 V2*. These fins would have been effective, yes, it was possible. But he eventually thought that all his flying machines needed vertical stabilizers/rudders out on the wing tip to control directional instability. So his competing flying machine to replace our all-wing *Horten Ho 229 V6* would have been a tailless one!

Gothaer Waggonfabrik started working on the first pre-production *Horten Ho 229*, known as the *Ho 229 V3*, in Autumn 1944. I know that the American Army troops discovered the *Horten Ho 229 V3* during the first week of April 1945. It was after bitter fighting in and around Friedrichroda with the *Waffen SS*, and the machine was not yet flight ready. But it should have been. All these pre-production machines had finishing dates established by the *RLM*. For example, the *Horten Ho 229 V3* was to have been flight ready in April, the *Horten Ho 229 V4* in May, and so on. Later, by August 1945, all the pre-production *Horten Ho 229 "V"* versions were to have been completed and series production started.

Commentary
David Myhra - Could the *Gothaer Go P.60B* have killed off the *Horten Ho 229 V6*? This is doubtful given all the political and *RLM* support the *Horten* brothers had acquired. This support included *Hermann Göring*, *Erhart Milch*, *Oberst Siegfried Knemeyer*, and *Oberst Artur Eschenauer* of the *Luftwaffe Quartermasters Office*,

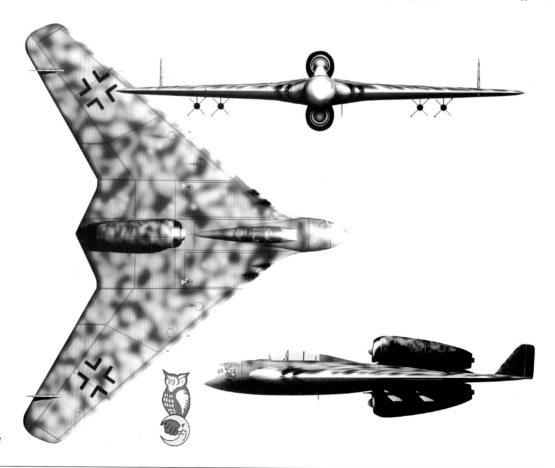

A three view drawing of the *Gothaer Go P.60C*.

A direct nose-on view of *Dr. Rudolf Göthert's* tailless heavy fighter *Gothaer Go P.60C*. Hidden are the navigator and radar operator due to their semi-prone positions under glazed hatches to the right and left of the pilot. The flying machine is shown with four *Fritz X* air-to-air wire guided missiles. Digital image by *Andreas Otte*.

and others. In addition, two versions had already flown: the *Horten Ho 9 V1* sailplane and the twin turbojet powered *Horten Ho 9 V2*. But in the end the *Hortens* and their *Horten Ho 229 V6* would have prevailed over *Rudolf Göthert* and all his academic colleagues from the several German aviation research centers. Yet the competition from *Rudolf Göthert* and his colleagues from these state-owned aviation research centers would have been beneficial in probably forcing *Reimar Horten* to seriously consider installing some form of a vertical stabilizer on their *Horten Ho 229 V6* with a hinged rudder attached. *Walter Horten* was ready to give it a try on their *Horten Ho 9 V2*, but after February 18th, 1945, it was no longer practical. *Walter* told this author that he would have placed a vertical stabilizer on their *Horten Ho 229 V3* sometime during its flight testing. *Reimar*, on the other hand, would have strongly resisted any such vertical surface. Nevertheless, there are people who believed that the turbojet powered all-wing was unstable at any speed...just as with *Jack Northrop's* all-wing *XYB-49*. Finally, critics claimed that the *Horten* brothers had no real experience with heavy, high wing loaded twin-engine fighters. Perhaps. However, the *Hortens* did have considerable experience with powered sailplanes...both single and twin engine versions. The *Hortens* lost their *Horten Ho 9 V2* during its flight testing. Who can say that it was lost, along with its pilot *Erwin Ziller*, due to the *Horten's* lack of experience in designing and building twin engine all-wing fighter prototypes? Then again, *Jack Northrop* lost two of his giant *XYB-49s* during their flight testing program.

The *Gothaer Go P.60C* would have weighed in at 23,082 pounds [10,470 kilograms] fully loaded with a maximum endurance of nearly 3 hours. Fire power would have included 4x*MK 108 30mm* cannon in the center section nose, as well as two oblique upward firing *MK 108 30mm* or 2x*MK 213* cannon in the center section. Digital image by *Andreas Otte*.

Port side view of the proposed tailless *Gothaer Go P.60C* three man heavy night fighter. Its overall performance would not have differed much from the all-wing *Horten Ho 229*, the machine it was supposed to replace. The *Go P.60C* was to have been powered by twin *Heinkel-Hirth HeS 011A* turbojet engines producing 2,866 pounds static thrust each. However, the *Horten Ho 229*, once well established in series production, was scheduled to receive the equally powerful *Junkers Jumo 004E*. So why replace the *Horten Ho 229* with the *Go P.60B*? *Dr. Rudolf Göthert* claimed that his tailless heavy day fighter design would be more stable than the all-wing *Horten Ho 229*. Digital image by *Andreas Otte*.

Huib Ottens.

Huib Ottens - During my visit to the National Space and Aeronautics Museum (*NASM*) in 1993, *Russell Lee* made available to me several documents regarding the *Gothaer Go P.60* verses the *Horten Ho 229*. They included:

1.0 - *Gegenueberstellung 8-229/Go P.60* - *Gothaer Waggonfabrik AG*, February 8th, 1945:

2.0 - *Projektvergleich der Jägervariante 8-229 und Go P.60* - DVL - February 16th, 1945;

3.0 - *Stellungname zum Projekt "Gothaer P.60"* - Horten brothers - February 16th, 1945;

4.0 - *Einige bemerkungen zur stellunahme von Horten zum Projekt "Gothaer P-60" vom 16.2.45* - Gothaer Waggonfabrik AG - February 22nd, 1945;

Document #2.0, in my opinion, is the most important of the four, because this was probably an official document published by a special evaluation committee of *DVL* that could have been used to influence the *RLM* in its choice between the *Horten Ho 229* and the *Gothaer Go P.60*.

There are notes in document #2.0 of a meeting to discuss the *Horten Ho 229* and the *Gothaer Go P.60*, but neither *Walter* or *Reimar Horten* attended this meeting, although *DVL* had invited them. The printed results of *DVL's* meeting showed that *DVL* was very much in favor of the *Gothaer Go P.60* tailless machine, and go on to repeat the statements already made in document #1.0. Documents #3.0 and 4.0 discuss the pros and cons of each other's flying machine.

Could the *Horten* brothers, by not attending *DVL's* meeting, have killed off their *Horten Ho 229*, thereby handing over the initiative to *Gothaer Waggonfabrik* and leading to Document #3.0, which they published after *DVL's* meeting? Their argument does not appear convincing enough, and it is my opinion that *Gothaer*

Electronic equipment on the proposed tailless *Gothaer Go P.60C* would have included:
- *FuG 280* - *FuG 101* - *FuG 130* - *EiV 125*
- *FuG 355* - *FuG 139* - *FuG 25a*
- *FuG 244* - *FuG 120K* - *FuBI 3*
Digital image by *Andreas Otte*.

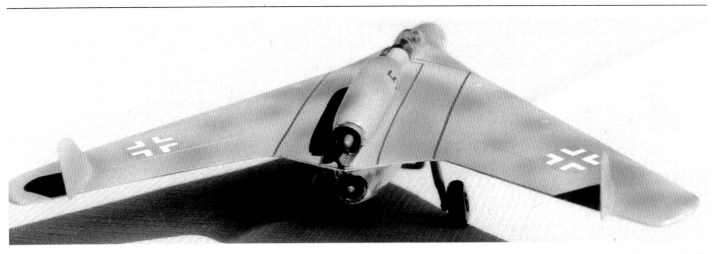

The trailing edges of the proposed tailless *Gothaer Go P.60* heavy night fighter. *Dr. Rudolf Göthert* placed one *Heinkel-Hirth HeS 011A* turbojet engine above and beneath the center section. He told *AAAFI* people post war that he disliked long ducted air intakes and exhausts, and therefore wanted the turbojets out in the open. This made them more easy to service. Scale model and photographed by *Reinhard Roeser.*

Waggonfabrik could have easily countered all of the *Horten's* points. Then again, whether both firms were using valid arguments is hard to establish after all these years. In addition, I do not have a background in aerodynamics and aircraft design. Nonetheless, both parties tried to convince the other one by claiming to have used the latest information and data base in designing their flying machines.

Based on the fact that the *Horten* brothers brought forward arguments based on their theories and experience gathered from their many test flights with sailplanes, including their *Horten Ho 9V1* and *Horten Ho 9 V2*, I believe that the weight of the *Horten* brothers' comments was heavier than the purely wind tunnel theoretical arguments put forth by *Rudolf Göthert* of *Gothaer Waggonfabrik*, although the *Horten Ho 9 V2* had crashed on February 18th, 1945, killing its test pilot *Erwin Ziller.*

On the other hand, the *Horten Ho 9 V1*, as well as the *Horten Ho 9V2*, were certainly not perfect flying machines and probably not the most ideal fighter aircraft due to their directional instability and other shortcomings. Even *Walter Horten*, an experienced former fighter pilot with *JG26*, had his doubts about the usability of the *Horten Ho 229* in combat and whether or not young inexperienced *Luftwaffe* fighter pilots with limited flight training could handle it in combat situations. Why didn't the *Horten* brothers defend their *Horten Ho 229's* design at the *DVL* meeting? Sure, they had experienced the ridicule of those academics before with their all-wing designs. Perhaps the *Hortens* believed that defending their all-wing *Horten Ho 229*, too, was a lost cause among these individuals. But perhaps the main reason for the *Hortens* not attending *DVL's* meeting with all the biased academics was that *Hermann Göring* had ordered it into production—so case closed. But perhaps even more importantly, *Hermann Göring* had given the *Horten* brothers a totally new emergency project...the all-wing *Horten Ho 18 "Amerkia"* bomber.

The proposed Gothaer Go P.60C would have used a conventional tricycle landing gear, however, the nose gear and wheel were to be offset to port for better pilot comfort. Digital image by *Andreas Otte.*

Leutnant Rudolf Opitz.
Germany, early 1940s.

Rudolf "Rudy" Opitz - Comments made to author during a taped telephone conversation regarding the directional stability of the *Horten* all-wing flying machines compared to the same from *Jack Northrop*.

Opitz: The *Horten* brothers did not have any production facilities, or the background, experience, and expertise to manage mass production of many airplanes like *Gothaer Waggonfabrik*. The *Hortens* never had anything produced, except for some glider research work, and they never came out of that stage. So how can someone even think of giving a project to them, when the important thing now was mass production of high numbers of fighter aircraft to stop the Allied bombers? The *Hortens* would be total newcomers to mass production of their *Horten Ho 229*. This just would not make sense, and apparently the *RLM* viewed it this way, too.

Myhra: *Erhard Milch* at the *RLM* probably knew and understood this, too.

Opitz: About a year ago, I saw a TV documentary on *Jack Northrop* and his all-wing bomber, the *XYB-49*. I learned that then Secretary

of the Air Force *Stuart Symington* wanted to force *Jack Northrop* to merge with *Consolidated-Vultee*, just so the flying wing would be built in Fort Worth, Texas, *Symington's* home state. That information was so horrifying. I think that documentary showed how great aircraft ideas can be killed for petty personal or political reasons. It's also a horrifying idea.

Myhra: *Symington* said that he had no memory of making any such statement and strongly denied that he ever demanded that *Jack Northrop* share work and merge with *Consolidated-Vultee,* or else risk cancellation of his big all-wing bomber...the *XYB-49*.

Opitz: Well, this story about what *Symington* said at the meeting was not *Jack Northrop's* idea alone. At that same meeting were several *Northrop* associates, a test pilot, and so on. They couldn't all have the same bad memory of what was discussed. Although the troublesome flight characteristics of the *XYB-49* all wing bomber were not yet solved at the time, perhaps with more people working on the problems they might have been solved. Another thing is unusual about the *Northrop XYB-49* story. The fact that the Secretary of the Air Force ordered the destruction of all existing aircraft, more than a dozen of them, is completely horrifying. Plus, all the valuable design and development work up to that point was stopped dead. Now today, with the *Northrop B-2 "Stealth"* bomber, the all-wing seems to again be in favor. But to remember that all that work on the *XYB-49* had to be destroyed then...this shows some of the things involved in pettiness and political power plays.

Myhra: *Gothaer Waggonfabrik* appeared anxious to get involved in a power play with the *Horten* brothers in order to kill the *Horten Ho 229* and substitute their *Gothaer Go P.60*. Just as *Symington* became mad because *Jack Northrop* would not bow down to merging with *Consolidated-Vultee.*

Opitz: You see this happening from time to time in every country.

Myhra: Did you ever see the *Horten Ho 9 V2* up close...*Reimar Horten's* twin turbojet-powered all-wing design?

Opitz: No, but I knew the man who flew it, though. *Erwin Ziller* was a very good friend of mine. The *Hortens* were not professional

Dr. Rudolf Göthert, as well as his colleagues from *DVL* and *LFA*, claimed superior handling characteristics of the *Gothaer Go P.60C*...all based on wind tunnel tests. However, the *RLM* knew the flying characteristics of the *Horten Ho 9 V2*...which would likely be similar to the *Horten Ho 229*. Besides, *Walter Horten* himself had promised the *RLM* that if the *Horten Ho 229* experienced directional control challenges then they would place a vertical stabilizer on the machine with a hinged rudder. Scale model and photographed by *Reinhard Roeser*.

The Gothaer Go P.60C was projected to have a service ceiling of 43,600 feet [13,300 meters] and a maximum range of 1,367 miles [2,200 kilometers]. It would have tracked down Allied bombers at night with its *FuG 240* scanning radar. Digital image by *Andreas Otte*.

The tailless *Gothaer Go P.60C* heavy night fighter as seen from its nose port side. There was not a great deal of friendship between *Dr. Rudolf Göthert* and the *Horten* brothers. Especially so, since *Göthert* and colleagues from *DVL* and *LFA* were seeking to convince the *RLM* to dump the *Horten Ho 229* in favor of the *Gothaer Go P.60C*. Digital image by *Andreas Otte.*

people and were not even close to successfully developing something so advanced like that *Horten Ho 9 V2*, using their own test pilots. The best test pilots are professionals who know engineering and design principles and who can give designers valuable information in order to make changes. The *Horten's* own pilots were not experienced, professional test pilots. First, these pilots could not give *Reimar* information on how to improve the design. These men were mostly experienced in flying gliders and were merely happy to fly whatever they could.

Myhra: What kind of man was *Erwin Ziller* in terms of his own flight testing experience?

Opitz: *Ziller* also was basically a glider pilot. That's how he and the *Hortens* met. Before the war, he had flown the 3-engine *Junkers Ju 52* to tow gliders. I think that he flew very few other powered aircraft...perhaps the *Fw 190* and the *Bf 109*, but not in any kind of testing capacity for *Focke-Wulf* or *Messerschmitt*. He flew the *Junkers Ju 87 Stuka* a lot. But the total flying experience which he had among different kinds of aircraft was very limited. None of the flying he did brought him anywhere close to a working knowledge of aerodynamics, or the engineering aspects of aircraft construction. The *Hortens* and their test pilots were just not up to that. From *Ziller's* point of view, everything the *Hortens* did appeared to be rosy and fine. There was never any attempt to improve anything,

A port side view of DVL and LFA's Horten Ho 229 slayer...Göthert's Gothaer Go P.60C. Scale model and photographed by Reinhard Roeser.

Starboard side of the *Gothaer Go P.60C.* Aviation historians looking for detailed engineering drawings by *Dr. Rudolf Göthert* have been unable to locate them post war. Were they purposly destroyed? Scale model and photographed by *Reinhard Roeser.*

that is, *Reimar* apparently had difficulty accepting ideas which came from the outside. Yet the *Horten Ho 4A* all-wing sailplane was a good performing glider. Research on the *Horten* all-wings just got stagnant. Then after the war, *Reimar* built an advanced model for the Argentines, but it didn't go any further there than his designs did in Germany. The basic design of the *Horten Ho 229* probably had many problems. The *Hortens* should face up to this and quit saying that their pilots were not experienced enough to handle those types of aircraft. That is not true, for they had *Heinz Scheidhauer* there to fly all their aircraft. He was a very highly qualified and experienced glider pilot. If the *Horten* pilots in Argentina could not do the job, then *Scheidhauer* could certainly have done it himself. But it did not work out this way. As I've said, the *Hortens* depended on people (test pilots) whom they themselves chose...just mostly inexperienced men who wished the glory of flying something very different. But these men were not the type of test pilots who would be able to tell the *Hortens* how to improve the design.

Myhra: Did *Ziller* ever talk to you about the *Horten Ho 9 V2* and its flying characteristics?

Opitz: No, we were far apart when *Ziller* was testing that aircraft. I was at Peenemünde, and *Ziller* had gotten out of the cargo-glider business. I was 100% involved in other things, too, and I did not have a chance to see *Ziller* after that. The only time we worked together was in the assault gliders in attacking and taking *Fort Eden Emeal*, France.

Myhra: Both *Horten* brothers blame *Ziller* for the crash of their *Horten Ho 9 V2* prototype.

Opitz: Well, I blame the *Horten* brothers for not having the foresight to get people there who could have helped them...people who could have given them honest opinions and design suggestions. The *Horten* brothers' biggest drawback was not accepting any outside opinion.

The *Gothaer Go P.60C* was to be a bad weather and night fighter...a tailless design which had been crafted exclusively through wind tunnel testing. It was hoped by the leaders at *DVL* and *LFA* that this proposed machine would kill off the all-wing *Horten Ho 229*. Scale model and photographed by *Reinhard Roeser*.

Myhra: Then the *Horten* brothers and *Alexander Lippisch* were like similar cuts from the same cloth? Neither *Reimar Horten*, nor *Alexander Lippisch* accepted outside opinions very well.

Opitz: In a way, *Alexander Lippisch* was not the easiest person, either, but he would accept a new idea and would be receptive to making changes.

Myhra: It seems *Reimar Horten* appeared less inclined to accept change. He seems to have been more hard-nosed about accepting change.

In the *Gothaer Go P.60C* heavy night fighter both the pilot and radar operator sat upright in tandem in the pressurized cockpit cabin. Scale model and photographed by *Reinhard Roeser*.

In March 1945, *RLM's* crew specifications changed with the addition of a navigator, and *Dr. Rudolf Göthert* redesigned the *Go P.60C's* cockpit. He placed the navigator and radar operator on each side of the pilot...one to the right and one to the left. The two men were seated in a semi prone position under a glazed hatch/panel within the wing contours. Scale model and photographed by *Reinhard Roeser*.

The extended nose on the *Gothaer Go P.60C* was due to the internal *FuG 240* inception radar. Scale model and photographed by *Reinhard Roeser*.

Opitz: I can only say that I got that same impression from that fellow (*Peter Selinger*) who wrote the book "*Nurflügel*." Well, I also remember that he got information from sources who were not necessarily in agreement with *Reimar Horten*. Some of this information caused serious difficulties between *Reimar* and *Selinger*. In the case of that *Horten Ho 4A*, for instance, *Reimar Horten* claimed that the Americans changed the aerodynamics when the aircraft was tested at Mississippi State University. This, of course, is not completely true. That was one of their problems...that they would not even think of talking to someone else. This made it very hard for anybody to penetrate the *Hortens'* thinking and to convince them to change. Instead, they were so obsessed with their own ideas...these were the only things they wanted to do and the only things they

would do. In the end, this was the reason for the failure of their *Horten Ho 9 V2* project. *Erwin Ziller* probably was as ill-equipped a test pilot for high-speed test flying as you could find, as far as his background and experience goes. He had flown mostly low-speed gliders, as I said earlier, and very low-wing loaded types of powered aircraft, such as the *Junkers Ju 52/3m*. All of this makes a very big difference. If the wing-loading numbers get up to those like the *Horten Ho 9 V2* had and you're also flying at very high-speed, then *Ziller* would not have enough experience in all those things. But experience is just one consideration. Basically, the airplane also had several bad design problems, and once *Ziller* was in an approach to landing with a dead engine, the directional control of the aircraft became completely inadequate. With one engine out, it just wasn't all the pilot *Ziller's* fault here. It was the design, too.

Myhra: When I asked *Reimar* about the directional control problems of the *Horten Ho 9 V2*, he told me that *Scheidhauer* had flown the *Horten Ho 7*, which was a twin-engine all-wing aircraft, with one engine out.

Opitz: How much power did that *Horten Ho 7* have, *David*?

Myhra: Well, it had twin *Argus* piston engines of 240 horsepower each, and it flew at a relatively slow speed.

Opitz: Plus a low-wing loading, I suspect. You know we can't compare apples and oranges here.

Myhra: Well, *Reimar* had *Scheidhauer* gave a flying demonstration before *Hermann Göring*, in which a *Horten Ho 7* flew very well on a single engine. He told me that all his twin-engine aircraft were designed to fly on a single engine if necessary.

The twin Junkers Jumo 004B powered all-wing Horten Ho 229 featuring straight air intake openings, compared to the upswept intakes on the *Horten Ho 9 V2*. Scale model and photographed by *Reinhard Roeser*.

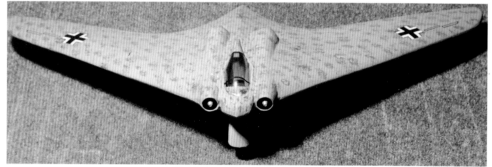

A ground level nose starboard view of the all-wing machine which created a great deal of anger and dislike at *DVL* and *LFA*...the *Horten Ho 229*. Scale model and photographed by *Reinhard Roeser*.

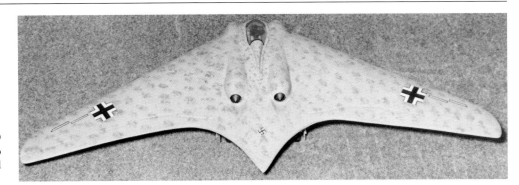

The trailing edge of the Horten Ho 229 featuring its buried twin Junkers Jumo 004Bs Scale model and photographed by Reinhard Roeser.

Opitz: That might very well be, *David*. But as I've said, to really compare the flight characteristics of different aircraft, you have to take into account what other people had to say who flew them. Those are the people who know. I can think of myself, and I'm a pretty good pilot, you know. I flew a number of the *Horten* aircraft, but I was never invited to fly a twin-engine *Horten*. I was going to fly the *Horten Ho 5C*, but several days before I was to fly it a professor from Göttingen University lost directional control at takeoff and crashed into the hangar roof, and it was no longer an aircraft available for me to fly.

Myhra: Were there any special difficulties in the flight path approaches at the Oranienburg Air Base?

Opitz: The Oranienburg airstrip was not so small as to cause the crash of the *Horten Ho 9 V2*. That's something else. That would not have made any difference at all. Despite this fact, the *Horten Ho 9 V2* was aerodynamically "dirty" with its landing gear out, and even though it was flying on just one engine, I understand *Ziller* "pancaked" in, so it means that he was too low. Any one of the old timers who flew *Horten* sailplanes, such as *Scheidhauer, Strebel,* and *Blech*, knew that the *Horten* all-wing flight characteristics were horrible. Yet somehow these aircraft flew. Even the *Horten Ho 4A* was a marginally performing aircraft...the flight characteristics were difficult and not acceptable. There was no doubt about it. It was always a funny thing. When I turned the *Horten Ho 4A* over to the Mississippi State University physics department, who bought it from me, I told them there were two things which they could do: You can clean it up and improve the performance; or, find a well-qualified pilot to fly it and he'll accept its poor performance. They wanted to

A nose port side view of a *Horten Ho 229 V3* with both of its *Jumo 004B* turbojet engine cowlings removed. *Dr. Rudolf Göthert* disliked the *Horten* brothers' practice of burying the turbojet engines. He preferred them out in the open. Unfortunately, we will never know, because the *Horten Ho 229 V3* was not completed by war's end, and the *Gothaer Go P.60C* was only a "paper project." Scale model and photographed by *Reinhard Roeser*.

enter it into the U.S. National Soaring Competition and just have something they could fly. But the American pilot who tried flying it quit the program, because he became absolutely horrified after flying it once. That's the same story that happened in Germany. Anyone who came from the outside and flew the *Horten* all-wing aircraft found that the aircraft flew, but that was about all. All these aircraft required considerable changes, but the *Hortens* would never listen. The *Horten Ho 4A* at Mississippi State University never went to any national competition because of its poor performance, despite the fact that they had good pilots there. It performed best only

A nose-on view of a *Horten Ho 229 V3* with its *Jumo 004B* turbojet engine cowlings removed for maintenance. Scale model and photographed by *Reinhard Roeser*.

The aerodynamically pleasing radar equipped Horten Ho 229 V6 bad weather and night fighter as seen from its starboard side. Scale model and photographed by *Reinhard Roeser*.

early in the morning, or in the evening...and then only a few test flights were made before that was the end of the test program. This tells exactly the story of the *Horten* all-wings. Their poor performance was not always due to the poor pilots. The men were not test pilots, like *Erwin Ziller*, and perhaps they did not fully understand the unusual behavior and aerodynamic characteristics of these aircraft. So most of them got scared out of flying these all-wings. I can personally talk about the *Horten Ho 4A* from here to there, because I have owned and flown it. But that doesn't help you, because everybody else thinks that it's a dog, or a lemon. This is exactly what happened here in America, as well as in Germany. You could never give the *Hortens* any advice, nor would they ask me, because you see I had worked once for *Alexander Lippisch*, flying the *Messerschmitt Me 163*. This sort of patronizing prettiness thing is what caused the difficulties. The British had that same *Horten Ho 4A* first; then the Americans had it. I flew it for them, and *Dr. Gus Raspet* from Mississippi State University was a world-wide

authority on aerodynamics....a very well-known and a very good name known the world over. He said "we'll just go for performance," but he could not really find people who could do it for him. He could find people who would fly it, but could not find anyone who could fly it, plus give an analysis or ideas on how to improve the aerodynamics and flight characteristics. This was the same way in Germany and Argentina. If the Argentine people were not up to flying the all-wing aircraft, why did *Reimar Horten* ever go there and choose them in the first place? I have different opinions. I knew the *Hortens* very well, and I have no reason to degrade what they did. But so far as their success goes, no matter how successful their designs were, success is not based on what they did inside their own company. It is based on the performance of their aircraft, and by their reputation in the world, just like anything else. The *Horten* all-wings just could not measure up to outside performance competition and critical aeronautical evaluations, so they can't blame anybody else but themselves for that. This they can't do. They had

The *Horten Ho 229*. Scale model and photographed by *Reinhard Roeser*.

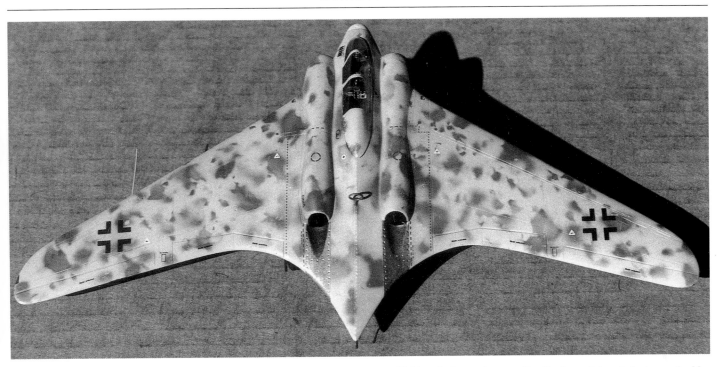

The Horten Ho 229 V6. Notice the coke bottle aerodynamic effect on the *Jumo 004B* turbojet engine nacelles. Scale model and photographed by *Reinhard Roeser*.

their chances. It is just like if I know somebody, and through their help, I get a good job. But once I'm on the job it is up to me to perform; it doesn't matter who I know. To get the job is one thing, and then I have to prove myself. Now the *Hortens* had many good opportunities. *Walter Horten* had very good connections with *Ernst Udet, Hermann Göring, Siegfried Knemeyer, Eschenauer, Todt*, and many others. He flew combat sorties with *Adolf Galland* during the Battle of Britain. The *Horten* brothers succeeded in using their contacts to get their research money and have production facilities set up. Now that was their chance to get their all-wing aircraft designs to succeed. But what they had to do was to prove the correctness of their ideas, and they were unable to succeed in doing that. This is the way it was. Just like that. The *Horten* planes were all basically fine. Their performance was basically fine, too. But if no others can fly the aircraft like they do...well, what does that tell you? Basi-

cally, that was the situation with the *Northrop* aircraft, too. It's not that their performance values were not correct. That was all acceptable. But an aircraft has to be more than just "flyable." I'm sure some of the aerodynamic problems could have been solved, they all have had their chances. The failure basically came because the flight characteristics were not acceptable. The same thing happened to the Russians when they flew with rockets. They canceled that research program because they killed a bunch of pilots. And this is the way it goes.

Myhra: Recently *Max Stanley*, a test pilot who flew the *Northrop* all-wings, gave a lecture at the National Air and Space Museum in Washington, DC, and afterward I asked him about the directional stability of the giant all-wing with only one engine operating. He told me that once when they were test flying the *XYB-49* all-wing bomber and returning home to Texas from Washington after a demonstration, they had to shut down three of the six turbojets because the ground-crew had failed to check the oil level in the engines prior to takeoff. Two of the failed turbojets were on one side, and *Stanley* said that the *XYB-49* flew just as well with two engines out on that side. He said that during his entire experience in flying the twin-engine *Northrop* all-wings, directional stability was never a problem with one engine out. So I asked *Stanley* to speculate on the stability of the *Horten Ho 9 V2* with one turbojet engine out, as it was coming in for a landing, with its high wing loading, and dirty? Well, *Max Stanley* is quite an advocate of the all-wing anyway, and it was his belief that most all-wings could be flown quite well on one engine.

Opitz: Oh, I don't doubt that it could be flown. *Mr. Stanley* speaks from a point of view of experience. The *Horten Ho 4A* could be

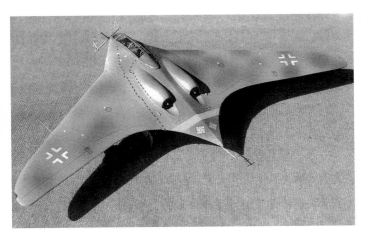

The Horten Ho 229. Scale model and photographed by Reinhard Roeser.

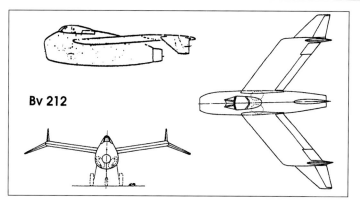

A competitor design to the *Gothaer Go P.60B*, the *Blohn und Voss Bv P.212* single seat and single turbojet powered day fighter design. *Blohm und Voss* was reportedly invited by the *Chef Technische Luftrüstung* on February 23rd, 1945, to construct three prototypes for tests and evaluation.

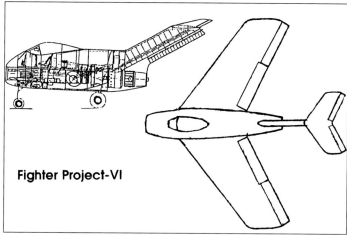

A competitor design to the Gothaer Go P.60B, the Focke-Wulf's Fw Ta 183 single seat and single engine day fighter design by Hans Multhopp. It is also known as the *Focke-Wulf Project #1*. It was reportedly ordered into production in April 1945 shortly before Germany surrendered...May 8th, 1945.

flown, too, but, nevertheless, its directional stability was marginal. Okay? When I flew the *Horten Ho 4A*, I had no real problem, but you always were flying this airplane in a marginal condition....so marginal, in fact, that it would be unacceptable to most pilots. Especially in light of the fact that it had no vertical control surfaces whatsoever, and it just had wing brakes near the wing tips. These so-called drag brakes made handling very easy at high speed, but were ineffective and very sloppy at low speed. So there it is. It all depends...if you recognize a problem, then you probably can get away with it....to a certain extent. I would not even want to talk about *Northrop's* big all-wing bomber aircraft. But one thing seems to be sure: *Northrop's* engineers never quite solved the problematic flight characteristics in the whole. They again hoped to solve these flight control problems by building bigger, with more power, and

adding more engineering and structural innovations to help overcome the negative flight characteristics. But flight control characteristics become more complex generally, not only in directional stability. So of course *Northrop* had a problem, too. But they even acknowledged that they had a problem...unlike *Reimar Horten*. I remember the time when they flew the *Northrop XYB-49* to Washington, DC, from California, and on the way home they landed at Wright-Patterson. I was working at Wright Patterson. But even before that I had talked to the *Northrop* people about the flight characteristics of all-wing aircraft. So I knew about its flight control problems. *Northrop* never denied this, and I'm sure they truly believed it could be solved in time with better control equipment. You know that today, if certain aircraft like *Northrop's B-2* did not have all the automated flight control equipment in them, they could not

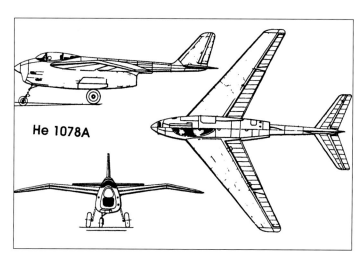

A competitor design to the Gothaer Go P.60B, the Heinkel He P.1078A. A single seat day fighter to be powered by a single Heinkel-Hirth HeS 011A-1 2,866 pound static thrust turbojet engine. It featured a 40 degree sweep back gull wing.

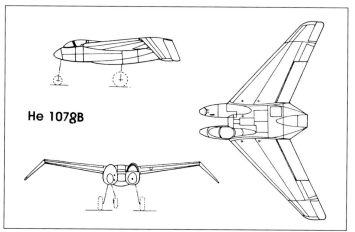

A competitor design to the Gothaer Go P.60B, the Heinkel He P.1078B of January 1945. An entirely new and different day fighter from the Heinkel He P.1078A and anything else up to that time and since. Although similarly powered as its sister *Heinkel He P.1078A*, this proposed machine would have been tailless, however, incorporating two small fuselage nose pods. The starboard pod contained the armament and nose wheel, while the port side pod contained the cockpit.

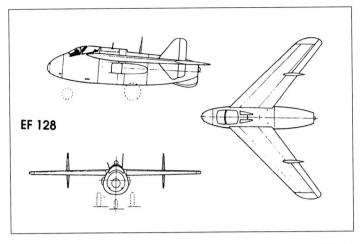

A competitor design to the Gothaer Go P.60B, the Junkers Ju EF 128 of October 1944. Design work on this proposed single seat day fighter with a single *Heinkel-Hirth HeS 011A-1* turbojet engine continued on through April 1945. Reportedly, a wooden mockup had been completed prior to war's end on May 8th, 1945.

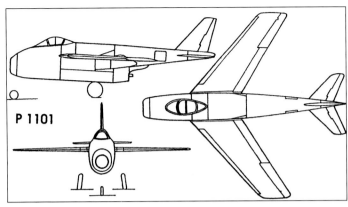

A competitor design to the Gothaer Go P.60B, the Messerschmitt Me P.1101. Rejected by the RLM, this single seat day fighter with a single *Heinkel-Hirth HeS 011A-1* turbojet jet engine and swing wing was found unsuitable due to the impossibilities of carrying weapons beneath its wings. Endurance and range was also limited due to the lack of sufficient fuel-carrying space.

be flown. So that's something else; a lot of progress has been made, and I am quite sure that the equipment available today would have solved the flight characteristic problems for both the *Hortens* and *Northrop*.

Goeff Steele - All these comparison books mentioned by *Huib Ottens* are all very interesting. Too bad there's not more actual writings from that era discussing that particular situation, for example, individual's memorandums, personal journals, and so on. I suspect—knowing a bit about the German culture and personality traits of these people—that this might have been a matter of professional jealousy between the established group of aerodynamicists and *Reimar Horten*, that young upstart who was trying to push his flying wing theories on everyone. *Reimar* was like an irritating horsefly, with a nasty bite thanks to *Walter's* position in the *RLM's* bureaucracy and his ability get public resources to support *Reimar's* playing with all-wing designs. So *Reimar*, that irritating horsefly, was too quick to slap down and kill. The established group of designers must have known or figured out how the *Hortens* were able to get money to continue their research and development of the *Horten Ho 9V2*, without some kind of factories and the resources available to *Messerschmitt AG*, *Heinkel AG*, and so on. They were probably jealous that *Reimar* had an inside connection via his brother *Walter*, and therefore some power and influence with the policymakers, and thus the *RLM* order to *Gothaer Waggonfabrik AG* to build the airplane in series production. They probably wanted to stop these outsiders, and particularly to put *Reimar* in his place, forcing more respect for authority...theirs. This sort of regimentation was a big thing in that era.

The culture of that time would have been to offer *Reimar*, the young upstart, a job somewhere after college graduation and absorb him quietly into the system, working with *Messerschmitt AG* or *Heinkel AG*, and allow him to learn and grow, under their proper tutelage, of course, until he could become a designer with a portfolio and a reputation. *Reimar* pissed all over *Heinkel's* patronizing

offer of employment and made demands *Professor Heinkel* could not ever hope to support. If *Reimar* went to *Messerschmitt AG* to ask for a job, I'm sure his reputation preceeded him. I suspect *Ernst* and *Willy* may have occasionally shared a beer together. The older generation would be particularly disdainful of someone like *Reimar*, who seemed to come from a liberal family and who had no apparent respect for authority. *Reimar* and *Walter* respected pure power like the *SS*, but seemed to have little tolerance for the respected authority figures, such as the aerodynamicists of the day, with the possible exception of his kindred spirit and competitor, *Alexander Lippisch*. Kids haven't changed a bit today, have they?

While the jury might forever be out on whether the *Horten Ho 229* design, or the *Gothaer Go P.60*, was actually a good, stable design, or a bad, unstable airplane not worthy of series production,

Geoff Steele.

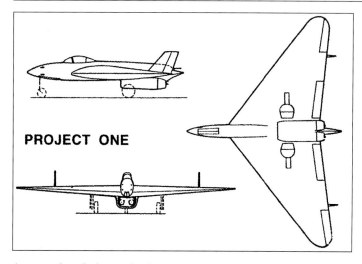

PROJECT ONE

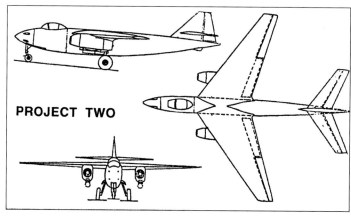

PROJECT TWO

A competitor design to the Gothaer Go P.60B, the Arado Projekt 1. A tailless delta-like night and bad weather fighter. Powered by twin Heinkel-*Hirth HeS 011A-1* turbojet engines, maximum speed of 503 miles per hour at 29,700 feet altitude for this 14-ton flying machine.

A competitor design to the Gothaer Go P.60B, the Arado Projekt 2. A design for an all weather fighter which was more conventional that its sister *Projekt 1* design. It would have been powered by twin *Heinkel-Hirth HeS 011A-1* turbojet engines mounted in a fashion similar to the *Arado Ar 234B*. Maximum speed was an estimated 481 miles per hour for this 29,820 pound flying machine.

there's no doubt the politics of the times were rife with discord. I'm betting that things were somewhat parallel in Germany and the United States, where our leading designers were trying to move ahead with advanced concepts for more traditional tube and wing aircraft, for example, fuselage with standard empennage of rudder and horizontal stabilizers. They came up against *Jack Northrop* and his flying wing theories. As in Germany with *Reimar*, *Northrup* had just enough influence in realms of power (Congress and the U.S. Air Force) to become an annoying pest. *Northrop* had his *XYB-49* all-wing design (to be built in California) approved for production of 12 prototypes (all these later designed to be converted to turbojet power), to be used as a fly-off competitor against the *Consolidated-Vultee (Convair) B-36* (to be built in Fort Worth, Texas), to see which airplane would get the U.S. Air Force strategic bomber contract for 200 or more aircraft.

On paper, the *Northrop XYB-49* was a great airplane and had specifications that outdid the *B-36* design. But in actuality, the flying wing just did not perform as promised. The *B-36* also had problems with its rearward facing engines, which would overheat and sometimes catch fire while driving contra-rotating propellers. I think that it was *Stuart Symington* (1901-1988), the Secretary of the Air Force who, conniving with one of the U.S. Senators from Texas, who got the *Northrop* design canceled, claiming it was a bad, unsafe airplane. The most insane part of the story was that the U.S. Air Force, actually *Symington*, issued an order that all the prototype *XYB-49*s were to be totally demolished and cut up for scrap, thus saving none of the airplane for future museums and so on. That was a deliberate, forceful slap in the face against *Jack Northrop* to leave nothing but a memory and a few photographs from his effort to have his airplane built. Most folks today say that it was

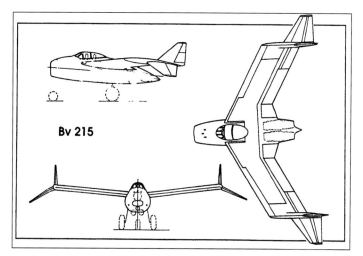

Bv 215

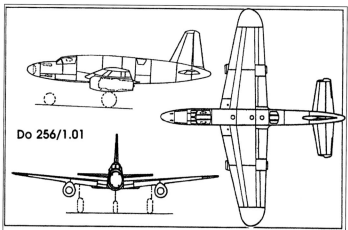

Do 256/1.01

A competitor design to the Gothaer Go P.60B, the tailless Blohm und Voss Bv P.215 night and bad weather fighter. Crew of three. Powered by twin *Heinkel-Hirth HeS 011A-1* turbojet engines located aft in the center section, and maximum level speed was calculated at 594 miles per hour. Service ceiling was 41,700 feet altitude.

A competitor design to the Gothaer Go P.60B, the twin turbojet powered Dornier Do P.256 night fighter. This was a two seat twin turbojet engine powered flying machine of conventional design that evolved from the *Doriner Do 335* single seat twin piston motored fighter. Maximum level speed was an estimated 515 miles per hour at 19,700 feet. Service ceiling was to be 41,000 feet altitude.

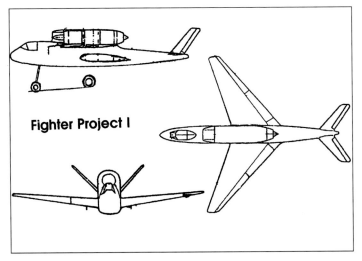

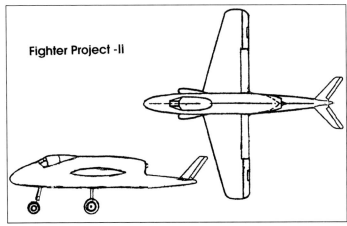

A competitor design to the *Gothaer Go P.60B*, the *Focke-Wulf Fw Proposal #1* day fighter of December 7th, 1942. A single seat and single *Junkers Jumo 004B* or *BMW 003A* turbojet powered flying machine with an unorthodox swept forward wing of 30 degrees. This is the first time a fighter design boasting a swept forward design appears in German aircraft project designs.

A competitor design to the *Gothaer Go P.60B*, the *Focke-Wulf Fw Proposal #2* day fighter of December 22nd, 1942. A single seat and single *Junkers Jumo 004B* turbojet engine with a straight wing and a swept back butterfly tailplane.

pure politics...a desire to keep post World War II weapons contracts distributed, instead of them all going to California...and that Texas deserved its share of the wealth. Whether the *Northrop XYB-49* was truly a bad airplane remains in conjecture. It never flew enough hours to really know; but it did kill a couple of pilots in its teething process when one *XYB-49* spun in, apparently uncontrollable.

There were other powerful characters involved in the death of the *XYB-49*. One was *General Curtiss LeMay*, and he may have had a lot of influence in really killing the giant all-wing bomber. *LeMay* had spent time with the Air Force test pilot *Cardenas* who had flown the *XYB-49*. It was *Cardenas* who told *LeMay* that the flying machine was unstable on its yaw axis and therefore unsuitable for precision bombing. That did not sit well at all with *LeMay*, whose *Boeing B-17s* could hit German targets, such as bridges and rail yards, with fair precision from 30,000 feet altitude. Though the *B-36*, by comparison, was stable because of its more traditional tube and wing design, and its huge size, too, it did have a tendency for its piston pusher engines to overheat and catch fire in flight. It was apparently a real handful to fly, as well, with a total of six reciprocating pusher engines running contra-rotating propellers with poorly designed gearboxes supplied by the U.S. government and four turbojet engines with two to a pod on each wingtip. Sort of like herding a big bunch of mechanical cats.

I suspect the *Horten Ho 229* was indeed unstable, though I would not necessarily call it a bad design. When its pilot *Erwin Ziller* lost one of his turbojet engines, the asymmetric thrust put the aircraft immediately into an envelope of marginal controllability, because there were insufficient lateral control surfaces and insufficient power/thrust to dampen adverse yaw and keep the flying machine flying straight and level. Normally, these are standard growing pains for any new design. Evolutionary changes to the airframe and control surfaces, plus development of more powerful engines probably could have rescued the *Horten* version of this design from

ruin. But I suspect the word may have gotten around after *Ziller's* crash and death on February 18th, 1945, that this was an unstable airplane and that it would be a pilot-killer.

The top management of the *Luftwaffe* certainly wanted the advantages offered by use of the turbojet engine, such as speed, altitude, and armament, and I suspect they secretly wanted to put all their emphasis into construction of the twin turbojet powered *Messerschmit Me 262*, which already was in series production and being flown with success against the Allied bomber formations,

Author *Dr. David Myhra*.

A ground level nose starboard side view of a tailless *Gothaer Go 60B*. The several pilots, no doubt, are describing their successes and failures in attacking Allied bombers with their new day fighter. Digital image by *Andreas Otte*.

mostly because of its speed. What was needed to perfect the *Horten Ho 229* were more advanced engines and better cannons—both under development, like the combination *Helmuth Zborowski* rocket motor and *BMW* turbojet, for momentary speed bursts and changes in the control surface geometry to make the airplane more stable. Perhaps it needed the fitting of a vertical stabilizer with an attached hinged rudder as *Walter Horten* wanted to placate the "old school of designers" and get the needed money from the *RLM*. The *Luftwaffe* had so few good pilots left that it needed a stable, proven airplane for them to perform their best, such as with the *Messerschmitt Me 262*. You just could not put unseasoned kids into a marginally stable aircraft such as the *Horten Ho 229* and expect them to master its intricacies without a lot of experience. *Luftwaffe*

leadership was not stupid, although *Adolf Hitler* sure was, trying to make the *Messerschmitt Me 262* a fast bomber and thereby delaying its deployment as a high altitude interceptor.

To have the *Hortens* dribbling away needed resources, such as turbojet engines, high-temperature metals, and so forth on a marginal design was thought to be not traditional and perhaps truly unstable by many pilots. Since it was therefore either very sensitive or impossible to fly safely, it may have seemed irrationally stupid to many *Luftwaffe* senior officers and some *RLM* management in the closing days of the war. That, notwithstanding the *Horten's* successes with their other sailplanes, particularly the *Horten Ho 4A*, which should have carried over into the *Horten Ho 229* design. But everything *Reimar* was building was really a one off experiment

A poor quality photo, deliberately rendered to represent war time conditions, featuring a ground level rear port side view of a tailless Gothaer Go 60B. Its nose gear, like its predecessor the Horten Ho 229 which it replaced, elevates it high above the tarmac. Digital image by *Andreas Otte*.

Bare chested ground crew turn and watch an unusual tailless flying machine in this purposely rendered poor quality photo. It is the Gothaer Go 60B in camouflage making a low altitude banking turn several hundred feet above the air strip. Digital image by *Andreas Otte*.

based mostly on the relative success/failure of the airship immediately proceeding it, and was not based on long years of advanced heavy fighter design and experience. The *Horten Ho 229* was *Reimar's* first attempt at a truly heavy airplane, and he was having problems with it. *Reimar* did not know enough and lacked the experience in how heavy, powered aircraft handle. Even the *Focke-Wulf Fw 190*, which was produced in series, was a serious under performer for *Kurt Tank* with its own stability problems, and it was a traditional fighter design.

The senior war leadership in Germany at that point was reacting to everything, anyhow. There was no plan anymore. Everything was chaotic and up for grabs. The *RLM* was in disarray, and everyone was ordering everything to be built. Priorities were changing, and countermanding orders were issued for production from one week to the next. And *Reimar* certainly was dealing with a wealth of heavy fighter design experience. All he had experience designing were lightweight sailplanes—certainly a different genre, for sure. The other classic designers were objecting to *Reimar's* wing designs on aesthetic grounds because they were not traditional. The other manufacturers were objecting to *Reimar's* designs because strategic resources were becoming more scarce and lots of fly by night small concerns like the *Horten* brothers were perceived to be bleeding off what was desperately needed in the main war effort, and they also wanted their hands on all the money the *Hortens* were getting, I suspect.

All the flight envelopes on the *Horten Ho 229* design—like the centers of gravity and lift, the control surfaces' responses, and so on—were very tight, that is to say, there was not much margin for error, if anything. The loss of an engine in flight pushed the pilot outside the normally tight operational box of control. Don't forget that *Adolf Galland* had flown the *Messerschmitt Me 262* in combat and was just hyped up on turbojets, period. I suspect if he thought the *Horten's* design would be a fast, stable, high-altitude, hard to hit gun platform, he would have leaped on it in a flash. But *Galland* was not flying the prototype, either. If he had tested the prototype instead of *Erwin Ziller*, perhaps he would have canceled the machine himself as unstable and unfit for production, in favor of building more *Messerschmitt Me 262s*. Or *Galland* might have known more about the operation of the *Jumo 004B* engine from

Three camouflaged *Gothaer Go 60B* tailless day fighters on patrol. All three are equipped with four *Fritz X-4* air-to-air offensive missiles. Digital image by *Andreas Otte*.

A pen and ink illustration of the *Fritz X-4* air-to-air offensive missile.

A poor quality photo of a tailless Gothaer Go 60B seen from its rear in its open hangar. About mid-1945. Digital image by Andreas Otte.

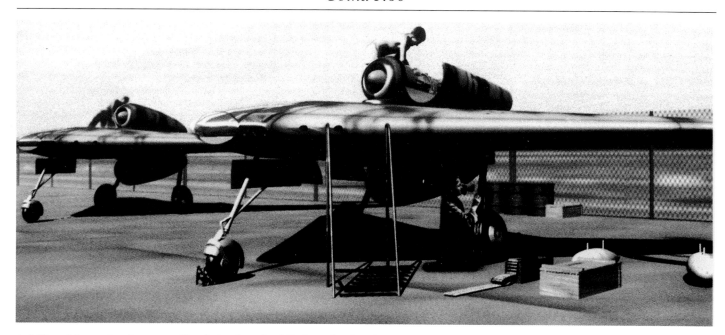

A deliberately rendered poor quality photo of a pair of *Gothaer Go 60As.* Both all-wing flying machines appear to be receiving maintenance to their *BMW 003A* turbojet engines. Digital image by *Andreas Otte.*

flying the *Me 262*, and could have saved the ship in a single engine flame-out situation. Who knows? I don't think *Erwin Ziller* was sufficiently experienced in flying turbojets, etiher, and I think that he knew it. In spite of the rush of turbojet-powered flying, I think *Ziller* was scared of the *Horten Ho 9 V2*, but felt under tremendous pressure to make it fly successfully, though perhaps he doubted he could. No one will ever know. I have not seen a copy of his logbook that his son, *Jorg Ziller,* has in Germany; perhaps that might contain some personal notes, such as the real capabilities of the *Horten Ho 9 V2* and his fear of flying it.

A purposely designed poor quality photo of a lone *Gothaer Go 60B* shortly after liftoff with dust being stirred up. The tailless *Go 60B* was to have been powered by twin *Heinkel-Hirth HeS 011A* turbojet engines producing 2,866 pounds of static thrust each. Digital image by *Andreas Otte.*

Five all-wing *Gothaer Go 60As* going on patrol. They would have been fitted with *4xMK 108 30mm* cannon and 4x *Fritz X-4* air-to-air guided missiles. Digital image by *Andreas Otte*.

The *Horten* brothers shown in an open auto...*Reimar* on the left and *Walter* on the right. No doubt about it...they had friends and support in very high places within the *RLM* and throughout *Nazi* Germany. A powerful force to be reckoned with.

Reimar Horten. He was the design genius behind the entire *Horten Flugzeugbau* family of all wing flying machines...both sail and power. Sensitive, shy, introverted, and reclusive, he was blessed with the gift of total recall.

The *Horten* brothers' initial thinking on an all-wing twin turbojet powered fighter prototype of 1941, which would later come to be known as the *Horten Ho 9*, and later still...the *Horten Ho 229*.

Walter Horten. He could have written the book on how to gain friends and influence people, as well as the book on networking. He had a remarkable ability to be instantly liked by anyone he met and gain their support for his brother's all-wing flying machines. Near war's end the aero academics at *DVL* and *LFA* thought that they could kill the *Horten Ho 229* in favor of the proposed tailless *Gothaer Go P.60* through wind tunnel research. Perhaps the *Gothaer Go P.60* could have been superior to the all-wing *Horten Ho 229* in terms of directional stability, but wind tunnel research would not have overcome *Walter's* Germany-wide network of friends and supporters.

Academics in Germany's universities of aeronautics believed that since the *Horten* brothers had no college background or access to wind tunnels the success of their all-wing flying machines was mere luck. Shown is the twin piston motored all-wing *Horten Ho 7* experimental fighter prototype of 1942.

Reimar Horten was considering a turbojet powered all-wing fighter prototype as early as 1941 when *Walter* learned of *BMW's* and *Junkers Jumo's* practical research in turbojet engines. Shown is a proposed Horten all-wing turbojet powered flying machine intended to use *Bramo/ BMW's 3302* turbojet engine.

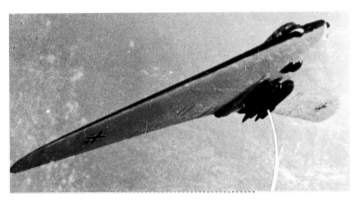

A poor quality photo of the bottom starboard side of the early *Bramo/ BMW 3302* turbojet engine powered *Horten Ho 9* of 1941.

A cut-away view of the *BMW 003A* powered *Horten* all-wing prototype day fighter of 1942.

Left: A photo of the *Horten Ho 9* of 1941 seen going away. German aeronautical teachers were largely unaware of the *Horten* brothers' practical research into all-wing flying machines. Since they had no access to *Nazi* Germany's wind tunnel network, they built one experimental flying machine after another in order to perfect the all-wing design.

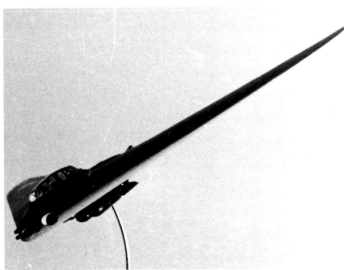

A port side view of the *Horten* brothers' initial turbojet powered prototype fighter from 1941.

The *Bramo/BMW 3302* turbine was canceled in 1942 for its failure to produce its design thrust. The *Horten* brothers were promised the next generation turbine from *BMW*...this was their *BMW 003A*, which had a static thrust of 1,760 pounds. *Reimar Horten* created a new design utilizing twin *BMW 003A* turbines as shown in this pen and ink illustration.

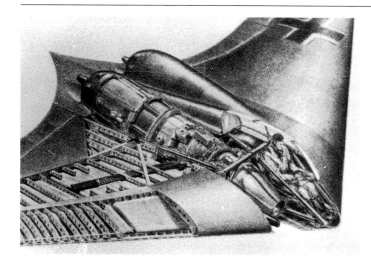

A close up of the early internal BMW 003A turbojet powered Horten all-wing, which was now being described as the Horten Ho 9. 1942/ 1943.

When *BMW* announced that their *003A* production would be delayed due to problems, the *Horten* brothers turned to *Junkers Jumo 004B* turbojet units for their fledgling *Horten Ho 9*. This pen and ink illustration shows the *Horten Ho 9* evolving into its highly aerodynamic shape, which would evolve further into the *Horten Ho 229*.

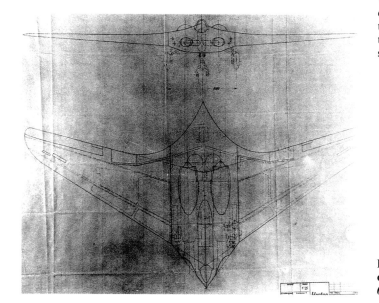

Left: A poor quality pen and ink drawing from the *Horten Flugzeugbau* of the maturing *Horten Ho 9* design, but still calling for twin *BMW 003A* turbines.

A ground level starboard side view of the *Horten Ho 9 V2's* metal tubular center section frame.

A view of the *Horten Ho 9 V2* rear featuring the jet nozzles of its twin *Jumo 004B* turbines. About mid-1944.

A poor quality photo of the center section leading edge air intake for the port side *Jumo 004B* turbine on the *Horten Ho 9 V2*, 1944. The air intake mounted *Reidel* turbine starter gasoline motor has not yet been installed.

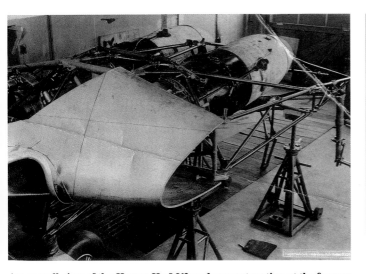

An overall view of the *Horten Ho 9 V2* under construction at the former *Autobahn* Vehicle Maintenance Center near Göttingen as seen from its nose port side, 1944. Certainly this engineering project was not the work of uneducated tinkers as the academics from *DVL* and *LFA* believed.

A rear view of the *Horten Ho 9 V2* project featuring its center section and the starboard outer wing, 1944. Plywood covering remains to be applied to the all-wing's center section. *Dr. Rudolf Göthert* would pretty much follow the overall design of the *Horten Ho 9 V2* when he designed his proposed *Gothaer Go P.60*...center section built out of metal tubular pipe, plywood covered center section, and removable outer wings built out of wood.

The *Horten Ho 9 V2* with its outer wings attached and nearing completion at its *Autobahn* Vehicle Maintenance Center near Göttingen as seen from its port side, 1944.

What appears to be a nearly completed *Horten Ho 9 V2* as viewed from its port side wing tip. Late 1944.

The *Horten Ho 9 V2* out on the tarmac at the Oranienburg *Luftwaffe* Air Base, northwest of Berlin, as seen from its port side. December 1944 to February 1945. The smooth lines of the center section were developed in the absence of wind tunnel testing, but were achieved through the systematic experimentation conducted by *Reimar Horten* over the years. Forty years later, *Northrop Aviation* aerodynamicists in the United States would return to the ideas of *Reimar Horten* in seeking to design a flying machine with the lowest possible radar cross section...the *Northrop B-2* bomber. The designs of *Rudolf Göthert* and his colleagues from *DVL* and *LFA* literally disappeared from the radar screen with the end of World War II.

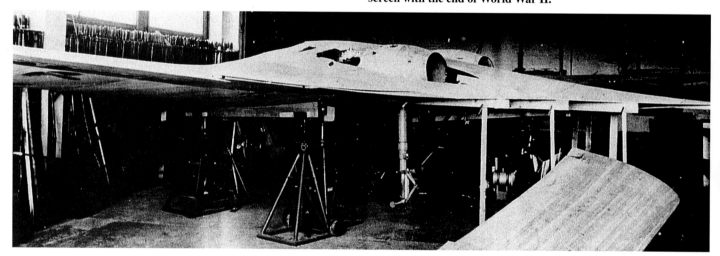

The nearly completed Horten Ho 9V2 as seen from its rear port side. Late 1944.

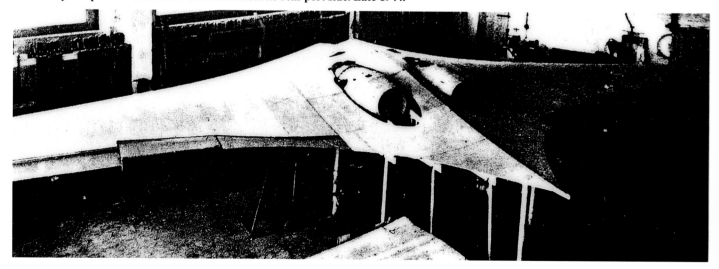

A view of the Horten Ho 9 V2 as seen from above. It appears complete save, for the metal cowling over the exhaust/thrust end of its twin *Jumo 004B* turbojet engines. Late 1944.

The Horten Ho 9 V2 on the tarmac at Oranienburg in late 1944 and early 1945. The all-wing Jumo 004B turbine powered flying machine is seen from its nose port side just prior to take-off. *Reimar Horten* sought to minimize any speed robbing items, so he placed everything he could inside the all-wing, a practice *Jack Northrop* had a passion for, too.

A close up of the *Horten Ho 9 V2* with its minimal radar cross section. This is where *Rudolf Göthert* of *Gothaer Waggonfabrik* and *Reimar Horten* differed considerably. Nevertheless, nice lines, however, a challenge overcoming directional stability in the absence of a vertical surface with an attached hinged rudder.

A poor quality photo featuring a ground level view of the *Horten Ho 9 V2*'s port side seen at Oranienburg. Its test pilot *Erwin Ziller* can be seen in the open cockpit. The *Horten* brothers joined exclusive company when it put its all-wing prototype fighter in the air. Only *Messerschmitt AG*, *Arado*, and *Heinkel AG* with their vast engineering resources were able to accomplish what the uneducated H*orten* brothers achieved with their *Horten Ho 9 V2*.

A ground level nose-on view of the minimal radar cross section all-wing *Horten Ho 9 V2* at Oranienburg. Compare this *Horten* flying machine to that of its competitor, the *Gothaer Go P.60B*.

Test pilot Leutnant Erwin Ziller, wearing his leather head cover, looks out of his Horten Ho 9 V2 cockpit at the camera man. Seen from the port side at Oranienburg November 1944 to February 1945.

A poor quality photo of the *Horten Ho 9 V2* as seen from its nose port side during flight testing at Oranienburg.

A poor quality photo of the Horten Ho 9 V2 during one of its take offs out of Oranienburg.

Horten Flugzeugbau test pilot *Leutnant Erwin Ziller* in the *Horten Ho 9 V2*. He died in a failed landing attempt in the *Horten Ho 9 V2* at Oranienburg on February 18th, 1945.

Left: A powerful supporter of the *Horten* brothers was the *RLM's* chief of Technical Development...*Oberst Siegfried Knemeyer* (far left). *Knemeyer* was also working with the *Hortens* in the development of the *Horten Ho 18B* "Amerika Bomber" project of late 1944 and early 1945. *Reimar Horten* is to the far right in the photograph.

The end of a test flight. The *Horten Ho 9 V2*, piloted by *Leutnant Ziller*, is photographed with landing gear extended while coming in for a landing at the *Luftwaffe's* Oranienburg Air Base, early 1945. It is difficult to see the flying machine from this angle due to its aerodynamic shape.

Digital rendering of an eye-witness account of how the *Horten Ho 9 V2* debris field looked immediately after the *Horten Ho 9 V2* spun in on the snow-covered farm field short of the runway at Oranienburg. Pilot *Ziller* was thrown into the fruit trees. Digital image by Mario Merino.

Reimar Horten in Argentina holding several scale models of his life's work: right hand includes the *Ho 3C*. In his left hand are scale models of the *Ho 4A* and *Ho 6*. 1982. Photo by author.

Reimar Horten checking out one of his all-wing experimental aircraft, 1943. He was 28 years old when this photo was taken.

Walter Horten. Baden Baden, West Germany, 1988. Photo by author.

Walter Horten. He is shown at the controls of an *Arado Ar 65* bi-wing fighter...the main aircraft of *I/JG 134 figh*ter/chase squadron based at Cologne. Mid 1936. He was 24 years old.

Right: A pair of piston motored all-wing Horten flying machines. Upper is the twin motored Horten Ho 5A from 1936-1937, while the lower machine is the Hor*ten Ho 2B* from 1935.

The all-wing *Horten Ho 1* of 1934. *Walter Horten* is seen in the sailplane's cockpit. *Walter Horten* test flew each of their experimental all-wing flying machines...except their Hor*ten Ho 9 V2*.

The highly successful all-wing sailplane from the *Horten Flugzeugbau*. This was their *Horten Ho 2* of 1935.

The single seat twin motored *Horten Ho 5C* of 1942. *Walter Horten* is seen standing (on a ladder) in front of the all-wing flying machine.

The motorized single seat *Horten Ho 3D* from 1941.

The two man individual cockpit twin motored *Horten Ho 5B* of 1937-1938.

A view of the twin motored tandem seat *Horten Ho 7* all-wing from 1942. This is the machine that *Oberst Siegfried Knemeyer* loved to fly, frequently driving out from the *RLM* in Berlin to Oranienburg in the northwest suburbs for a test flight...perhaps just to relax, or trying to figure out how to deal with the closing days of the *Luftwaffe*. *Reimar Horten* told this author that their *Horten Ho 7* was the most pleasing all-wing flying machine in their entire stable.

The single seat two motored *Horten Ho 5C* featuring its port side.

Walter Horten had connections. In this photo from mid-1941, *Walter Horten* (pilot far left) is escorting *Hermann Göring, Erhard Milch, Ernst Udet,* and *Field Marshal Albert Kesselring* to a *Junkers 52/3m* transport aircraft. All these top *Luftwaffe* personalities were trusting in *Walter Horten* to get them to their destination.

Reimar Horten is shown at a formal gala in Berlin in the late 1930s. From left to right: *Erhard Milch*, unknown Italian General, *Reimar Horten, Ernst Udet,* and *Ernst Heinkel.*

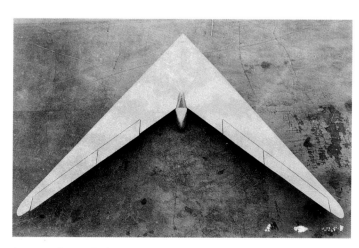

An experimental single seat research sailplane from 1944-1945 and known as the *Horten Ho 13A*. It featured a highly swept back wing of 60 degrees. This is only one example of how the H*orten* brothers carried on their aerodynamic research through the construction of sailplanes in the absence of wind tunnel access.

General Kurt Diesing... Hermann Göring's chief of staff in Berlin and supporter.

Reimar Horten is photographed speaking with *RLM's General Kurt von Döring*...head of the Inspection of Fighters Command, *Walter Horten's* boss at the time, and an avid supporter.

One of the *Horten* brothers' foremost supporters in seeing that they had facilities, engines, and supplies in constructing their all-wing experimental flying machines...the powerful and influential *Oberst Siegfried Knemeyer* from the *RLM*.

Left: *Oberst Autur Eschenauer* of the *Luftwaffe's Quartermaster's* Office in Berlin. *Walter* and *Reimar's* older brother *Wolfram* and *Eschenauer* were in pilot training school at the same time and close friends. *Wolfram* was lost off the coast of France in 1941 dropping sea mines from a *Heinkel He 111* when it suddenly blew up. *Eschenauer* befriended *Wolfram's* younger brothers in their attempts to perfect the all-wing planform with an unending supply chain from the *Quartermaster's* Office.

It was *Autur Eschenauer* who had assigned the *Horten Ho 229* to *Wolfgang Späte's JG400* fighter group, replacing their aging and largely ineffective *Messerschmitt Me 163* rocket powered interceptor.

JG400's coat of arms featuring the *Messerschmitt Me 163* rocket interceptor. *Eschenauer* had told its commander *Wolfgang Späte* that they would soon be receiving the all-wing *Horten Ho 229* under series construction by *Gothaer Waggonfabrik*. Would *Späte* have placed the *Horten Ho 229* over the *Messerschmitt Me 163* seen in the *JG400's* coat of arms?

Walter Horten, seen up on a ladder, is peering into the cockpit of their single seat twin piston motored *Horten Ho 5C.*

Reimar Horten is seen standing near the starboard outer wing of their *Horten Ho 9 V1* all-wing sailplane.

Walter Horten (2nd from the right) is shown with four of *Erhard Milch's* senior staff. The *Horten brothers* had the support of *Milch*, a powerful senior member of the *RLM*...that is, until *Hitler* fired him in late May 1944 for attempting to produce the *Messerschmitt Me 262* as a dedicated fighter instead of the blitz bomber which *Hitler* had demanded. This photo is from 1944.

General Hermann, chief of staff in *Erhard Milch's* Berlin office and a friend and frequent visitor to the *Horten* workshop where the *Horten Ho 9 V2* was under construction. Notice that in the background can be seen the rear port side of the *Horten Ho 9 V1* sailplane.

Left: What would the *Horten* brothers have done with the *Messerschmitt Me 262*? *Reimar* and *Walter* told this author that they did not want any part of developing this very good turbojet powered fighter because they were all-wing advocates. Development was not their interest...thus, their *Horten Ho 9 V2* went to *Gothaer Waggonfabrik* in late 1944 for further development and series construction as the *Horten Ho 229*.

Horten supporter *General Erhard Milch* of the *RLM*. In late May 1944, *Milch* was having his troubles with *Willy Messerschmitt*. It got to the point where *Milch* told *Messerschmitt* that he was going to give the development of his *Me 262* to the *Horten* brothers because they did what they promised. It did not happen, of course, but this shows the good reputation the *Hortens* had within the *RLM*.

Reichsmarschall Hermann Göring...the Horten brothers' biggest fan.

Willy Messerschmitt. Send his pride and joy to the all-wing *Horten* brothers? Impossible. *Willy Messerschmitt* was hardly on speaking terms to the much younger *Horten* brothers, and he came to hate them after *Milch* threatened him with the loss of his *Me 262*. The *Horten* brothers did not like being used this way, either.

Reimar Horten's Ho 18A "Amerika Bomber" project. *Göring* wanted revenge, while his chief planner, *Siegfried Knemeyer*, wanted a way to bring the destruction of Germany to a stop. The six *Jumo 004B* powered *Horten Ho 18A* would be used to make a round trip flight to New York City. Once the fast and high flying all-wing was over New York, it would drop a 5,000 pound high explosive conventional bomb. However, this bomb would have been wrapped with multiple blankets of radioactive silica. *Knemeyer* and others hoped that within a few days New Yorkers would become sick and die of radiation poisoning. Thus, *Knemeyer* believed that Germany could sue for a cease fire, or at least some form of conditional surrender. *Knemeyer* had instructed *Dr. Eugen Sänger* to design and construct his plans for an orbital bomber...to bomb America from space, as well as the *A-9/A-10* rocket missile by *Wernhen Von Braun*.

The new and improved *"Amerika Bomber,"* the *Horten Ho 18B*. It was smaller than the *Horten Ho 18A*, plus it would do just as well with only four *Heinkel-Hirth HeS 011A* turbojet engines with their 2,866 pounds of thrust each. This was the *Horten* brothers' final aircraft design for *Nazi* Germany.

Jägerstab Karl Otto Saur, from *Speer's* Ministry of War Production, was in charge of the new fighter production program, which also included the *Horten Ho 229 V3*. The *Ho 229* had been included in the fighter production program as a high priority item.

The *Horten Ho 18A* seen over New York City. Digital image by *Ken Spacey.*

Albert Speer (right), chief of the Ministry of War Production in early 1944. *Göring* had selected the *Hortens* to design and construct their *Horten Ho 18B* "Amerika Bomber" in late 1944 and as quickly as possible. *Speer* was making underground production facilities ready for the *Horten* brothers' *Ho 18B* in one of the largest underground factories...Kohnstein, near Nordhausen in the Harz Mountains. *Speer* would provide sufficiently large numbers of foreign workers and slave laborers to get the *Horten Ho 18B* built.

Jägerstab Karl Otto Saur was also making ready the new underground factory at Muhldorf with its 18 foot thick roof. Here, in this huge bomb-proof hangar, the *Horten* brothers could construct their *Horten Ho 18B* "*Amerika Bomber*" with inmates from nearby concentration camps (*KZ*).

Entrance to Jägerstab Saur's Muhldorf underground assembly facility. Saur was found guilty of crimes against humanity at Nuremburg and hanged.

Inside the huge underground assembly at Muhldorf. Notice the size of the men to the right of the photograph relative to the height of the structure's roof.

Autur Eschenauer from the former *Luftwaffe's Quartermaster General's* office. Photo taken post war.

ISNTF-1
By *B L Masterson*

COPY NO.

DS2.1
Horton /3

HEADQUARTERS
UNITED STATES STRATEGIC AIR FORCES IN EUROPE
Office of Asst. Chief of Staff A-2

AAF Station 379
APO 633, U S Army
12 May 1945.

TECHNICAL INTELLIGENCE)
REPORT NO.........I-4)

SUBJECT: Horton 229 Bat Wing Ship Construction.

1. Pursuant to instructions from Operations Officer, ATI-Roth, R-46 the undersigned investigated subject target this date and interrogated the German civilian Engineer detained therewith at target location. As a result the following preliminary report is submitted.

    a. Report

        (1) NAME OF TARGET: Horton 229 Bat Wing Ship Construction.

        (2) LOCATION: 45th ADG, code name-"Gunfire"- situated in town of Wolfgang, GSGS 4346, M-872-694 (near Hanau) (Airplane picked up by 45th ADG at Fredrichroda, R/5 090550).

        (3) DESCRIPTION: Twin jet propelled bat wing constructed airplane, tricycle landing gear. The airplane is of steel and plywood construction and this particular ship has never been flown. It is presently disassembled in three major sections namely the right and left wing sections, and the center section consisting of the nose-jet engines-landing gear. The center section cannot be disassembled because of welded construction.

            Size disassembled, center section--24' long, 10'-8" wide.
            Wing panels--12' wide, 28' long.
            Size assembled, wing span approximately 65', length 24'.
            Weight, approximately 5 ton.

        (4) INTERROGATION OF ENGINEER: Herr Eckhardt Kaufmann, civilian, resident of Fredrichroda, Germany claims to be the supervisor of construction and assembly of this airplane. He states that a Major Horton of the German Luftwaffe is the inventor and designer of this ship and that he is supposed (by Kaufmann) to be a P.W at present; that he was ordered to build three of such airplanes at the Gotha W.F. (Car Foundry) of which a Dr. Berthold is director; that the ship was designed as a fighter with provisions made in plans for wing guns; that its maximum speed was to be 900 km/hr; that one ship was built, test flown at 300 km/hr and landed safely but on the second flight, the left engine burned up and the ship crash landed. Upon further questioning Herr Kaufmann admitted that practically all the materials used in construction of this ship are too heavy and improper, and are not in keeping with materials currently used in aircraft construction but that aluminum, dural, etc. were not available and that he was to use whatever material he could find. He further claims that on or about March 15, 1945, General Young, in charge of jet airplanes, came to him and offered proper materials, stating that airplane production in Germany had stopped, but that it was too late because construction had progressed to it's present status. Herr Kaufmann claims that plans in our possession incorporate improvements over this airplane.

        (5) Herr Kaufmann has expressed a desire to work with American Engineers in America toward completion and perfection of this plane. He claims that with his knowledge and existing plans he could, with proper materials, build even a better plane without the use of the present prototype. Herr Kaufmann also expressed a desire that his family accompany him if he went elsewhere to work on these plans for fear of reprisals against them in Germany.

-1-

Organization   Go P-60

Interrogation of *Dipl.-Ing. Eckhardt Kaufmann*, supervisor of construction on the *Horten Ho 229* at *Gothaer Waggonfabrik, Friedrichroda*. May 1945.

TI Rpt. I-4, 12 May 1945

(6) REMARKS: In view of the foregoing it is doubtful whether completing and test flying this plane would give results that could be evaluated because of it's excessive weight resulting from use of wood and steel in construction. Due to the bulk and weight of the airplane it was deemed advisable to hold same at it's present location, pending evaluation of plans and further interrogation of Herr Kaufmann by ATI-USSTAF.

2. Herr Kaufmann and one steel locker containing plans for the Horton 229 airplane accompany this report in the custody of an officer guard from Headquarters, GUNFIRE.

3. One complete Horton 229 airplane as described herein and parts of another are being held at GUNFIRE by Plans and Operations Officer, pending further instructions from ATI.

Basic report prepared by Herbert A. Hazen, Maj., AC,
Approved by Mark Bradley, Col., AC, Executive.

H. D. Sheldon. Col AC
GD

for: GEORGE C. McDONALD
Brig. Gen., U.S.A.
Asst. Chief of Staff A-2

-2-

Classification Cancelled
or Changed to
Auth: *R&R 1/22/46*
TSNTE-1
By *B L Masters*

HEADQUARTERS
UNITED STATES STRATEGIC AIR FORCES IN EUROPE
Office of Asst. Chief of Staff A-2

COPY NO.

AAF Station 390
APO 633, U S Army
19 May 1945

TECHNICAL INTELLIGENCE)
REPORT NO.........I-3)

SUBJECT: Interrogation of Herr Eckhardt Kaufmann Concerning Ho 229 Aircraft.

1. Herr Eckhardt Kaufmann, who stated that he was supervisor of construction
and assembly of the Ho 229 aircraft for Gotha Waggenfabrik A.G. in the plant at
Friedrichroda (50°48'N/10°33'E), had been sent back to USSTAF (Main) by ATI personnel. He was interrogated 14 May 1945 at Chateau du Grand Chesnay, near Headquarters, SHAEF (Main), by Lt. Col. A. B. Deyarmond and Capt. S. Litton of the
Office of the Asst. Chief of Staff, A-2, USSTAF.

2. Digest of statements by Herr Kaufmann.

a. The Gotha Waggenfabrik A.G. of Gotha ordinarily builds railroad cars,
but during the war about half of their capacity was devoted to aircraft manufacture.
They built Me 110 aircraft in quantity, attaining a peak production of about 120
in September, 1944. They had started work on FW-152's and had partially constructed five (5) which were about ready for assembly. Because of lack of progress by
Horten in the development of the Ho 229 this job was turned over to Gotha W.F.
They had partially completed three and had a contract for another ten, after which
they were supposed to go into quantity production. This product was considered a
very high priority job and they all worked very hard and very long hours on it.

b. The key personnel of this company who were involved in the Ho 229 project were:

(1) Dr. Berthold, Director.

(2) Herr Huehnerjaeger, Chief Engineer. Under him were:

    (a) Herr Mueller in charge of control systems; formerly with
         Junkers.

    (b) Herr Schaupp in charge of power plant.

    (c) Herr Hermann in charge of body design.

    (d) Herr Freitag in charge of landing gear.

    (e) A fifth engineer, whose name Herr Kaufmann did not remember,
         was in charge of design of skin and covering.

    (f) About fifteen draftsmen.

(3) Herr Kaufmann was supervisor of construction and assembly and
     directly responsible to Dr. Berthold.

c. There were no resident government inspectors on the project but a
district inspector visited there periodically to check the parts being manufactured.

d. There was also one Aluis Boensch, a Junkers employee, who was a specialist on the engines. He was in the Army but placed in an inactive status in
order to do that work. When the district was taken by the Americans he was sent to
a P/W camp at Laucha. ATI personnel visited the location of this camp but it had
been moved and they could not trace Boensch.

e. Other companies associated with Gotha in the manufacture of this aircraft were:

    (1) Hartweg in Sonneberg (about 80 km. from Gotha) made the wings.

(2) Komprinz in Ludwigshafen made the landing gear shock struts.

f. The manufacture and testing of this aircraft was accomplished at several dispersals of the Gotha plant.

 (1) The Friedrichroda plant assembled the aircraft and made most of the major parts.

 (2) The plant at Luisenthal did the welding work on the body using parts furnished by Friedrichroda.

 (3) The small mechanical parts were to be built at a number of small dispersal plants near Friedrichroda.

 (4) The design and engineering department was at Ilmenau. At this place there was a mockup of the first of the production series (fourth aircraft built by Gotha) which was to embody all the improvements found necessary in the construction of the first three. Herr Kaufmann did not know whether or not this place was visited by ATI personnel.

 (5) Static testing was done at the testing laboratory at Wutha near Eisenach. They had complete testing facilities there where the Gotha W.F. accomplished tests under inspection by government representatives who were resident inspectors there.

g. The first prototype of this aircraft was designed and built by Horten and flight tested at Neu Brandenburg. They had one successful flight which indicated that the ship flew well but the landing gear was damaged in landing. In March, on the second flight, the left engine failed while coming in for the landing and the aircraft cracked up and was destroyed on landing. Meanwhile, the Government had turned this project over to Gotha W.F. for further development. They were constructing three of the aircraft, making necessary modifications as they went along. Their engineers had visited the Horten establishment, taken necessary measurements, made revisions in the drawings, etc. At the time the Friedrichsroda plant was captured, Gotha had completed the body construction on two aircraft. The wings for the first were nearly ready for assembly, but the wings for the second aircraft were just getting started. The third aircraft was incomplete. Two drums full of drawings were dug up by the ATI personnel and taken away. The fourth airplane was to be the first of the series of ten (10) for which a contract had already been awarded, and was in the mock-up stage. This aircraft was to embody all modifications and be the model for the production series. There were also plans to make it a two-place plane, although this does not show in the mock-up. All new design work on this model was to be done by Gotha W.F. and not by Horten. The landing gear retracting system was to be completely changed and this revision is shown in the mock-up. Armament, armor plate, and self-sealing fuel tanks were to be incorporated in the fourth aircraft.

h. If they had not been stopped by the Allied invasion, Herr Kaufmann believes they could have completed the order of ten (10) aircraft in about six months and could have been in full production in another six (6) months. They expected to manufacture about 1,000 aircraft a month of this aircraft, using several manufacturing plants.

i. Details of construction of this aircraft can be observed by inspection of the articles captured, but the following points were described:

 (1) The body structure was made of welded steel, described by Herr Kaufmann as 60 kg. chrome-steel. The significance of the figures could not be determined, although he said it was the Brinnel hardness of the steel. The only heat-treated parts he knew about were the engine-mounting bolts which he said were 90 kg. steel heat-treated to 120 kg.

B34 -2-

X 142 978

TI Rpt. I-3, 19 May 1945

(2) The covering of the body was plywood six mm. thick at the rear, twelve mm. in the middle and sixteen mm. at the nose. Behind the jets is a mild steel plate covering over the wood. Casein glue was used in joining the wood.

(3) The wings were made of wood. He did not know the details of construction as another company made these parts. However, he did say the spars were made of vertical laminations. He thought that fir, beech, and birch wood were used in various parts. The covering was plywood.

(4) The largest wheel of the tricycle landing gear is the nose wheel.

(5) There are three trailing edge moveable control surfaces on each wing. The two inboard acted as elevators and the outboard as ailerons and elevators. Spoiler flaps, top and bottom, at the wing tips were used for directional control. The details of use of controls could not be checked as Herr Kaufmann was not familiar with them. Lower surface flaps near the center could be used in landing or pulling out of a dive. A decceleration parachute was available for emergency use.

(6) The fuel tanks, four in each wing, had a total capacity of 3,000 liters. Herr Kaufmann thought the aircraft would have an endurance of three quarters of an hour to one hour at 20,000 to 25,000 feet.

(7) The armament was to be four Mk 108 or two Mk 103 cannons. The armor plate was to be 10 mm. thick.

(8) A catapult type seat was provided using an explosive propellant.

(9) The first ship was tail heavy and used about 800 kg. of nose ballast. Mr. Kaufmann thought the installation of armament and armor would make up for this. The ballast was 200 kg. in the very nose and 600 kg. in two places about 1600 mm. back. It could be replaced by a total of 438 kg. in the nose.

j. Herr Kaufmann knew nothing of aerodynamic details or of the strength design of the aircraft. He did not have with him a detail weight breakdown.

k. Herr Kaufmann brought with him a box of documents which did not seem to be very important. Some of the correspondence indicated that the government wanted this project rushed. This box of documents is being held by Exploitation Division, Office of the Assistant Chief of Staff, A-2, US Strategic Air Forces in Europe, pending further disposition. One interesting item was an illustrated parts list for the Jumo 004 engine.

3. Recommendations.

a. That the appropriate ATI Collection Point investigate the design and engineering establishment of Gotha W.F. at Ilmenau, inspect the mock-up and take necessary photographs.

b. That an attempt be made to find aerodynamic and performance reports on this aircraft. Copies may be at Ilmenau.

c. That engineering personnel of Gotha W.F. and of Horten be questioned if available, but not brought to USSTAF for this purpose.

d. That the static test laboratory at Wutha be inspected by ATI personnel from the appropriate Collection Point, to determine whether anything more of value can be learned.

TI Rpt. I-3, 19 May 1945

Basic report prepared by A.D.Deyarmond, Lt. Col., AC., S. Litton, Capt., AC.,
approved by Loyd K. Pepple, Col., Ord., Chief of Section.

GEORGE C. McDONALD
Brig. Gen., U.S.A.
Asst. Chief of Staff A-2

Left to right: Americans *General Carl Spaatz*, *Colonel Harold Watson*, and U.S. Air Force Intelligence chief...*General George C. McDonald*.

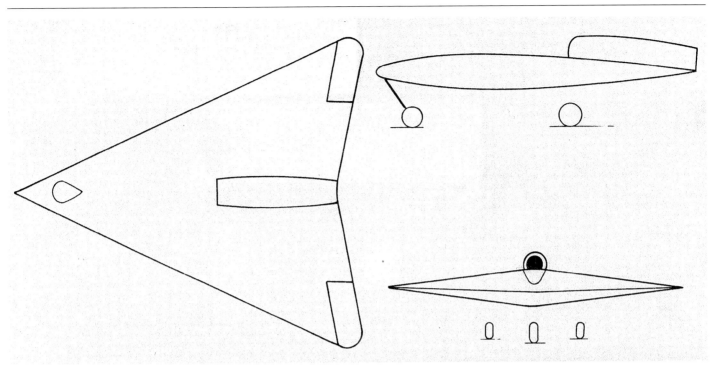

Above: A pen and ink drawing of the delta-winged and prone piloted *Horten Ho 10A* **featuring its single** *Heinkel-Hirth HeS 011A* **turbojet engine mounted over the aft fuselage.**

Right: A pair of single turbojet engine powered *Horten Ho 9Bs* **in formation. Digital image by** *Mario Merino*.

The proposed single seat, single *BMW 003A, Junkers Jumo 004B,* or *Heinkel-Hirth HeS 011A* turbojet engine powered *Horten Ho 9B*. Its turbojet would have been mounted over the upper surface of the fuselage and directly behind the cockpit. Scale model and photographed by *Reinhard Roeser*.

A direct nose-on view of the proposed dorsal-mounted single turbojet powered Horten Ho 9B. Digital image by Mario Merino.

The giant six *Junkers Jumo 004B* turbojet-powered all-wing *Horten Ho 18A "Amerika Bomber."* If the combined *Junkers/Messerschmitt* committee had had their way, the all-wing flying machine would have been equipped with a vertical stabilizer with a hinged rudder.

An airborne view of the *Junkers/Messerschmitt* committee redesigned *Horten Ho 18A.* Scale model and photographed by *Hans Peter Dabrowski.*

A port side profile of the proposed Horten Ho 9*B* with its single dorsal-mounted turbojet engine behind its single seat cockpit. Digital image by *Mario Merino.*

An overhead view of the combined *Junkers/Messerschmitt* committees' effort at designing an *"Amerika Bomber"* based on *Reimar Horten's Horten Ho 18A* in late 1944. After *Junkers/Messerschmitt* committee was thrown out by an angry *Hermann Göring*, *Junkers* renamed the project as the *Junkers' Ju EF P.140*. *Göring* personally saw to it that no one pushed *Reimar Horten* around. Scale model and photographed by *Hans Dabrowski*.

Hermann Göring seen at his huge desk top at *Karenhall* reviewing *Reimar Horten's* plans for the *Horten Ho 18 Amerika* Bomber project in late 1944. He told *Reimar* to forget about the *Junkers/Messerschmitt* committee-redesigned *Horten Ho 18A*. The *Horten* brothers had no greater supporter for their all-wing aircraft than this man...the chief of the *Luftwaffe*. Pretty much what *Reimar* wanted *Hermann Göring* gave, and with great generosity. No aircraft manufacturer, let alone a small firm such as *Gothaer Waggonfabrik*, would dare rename the all-wing *Horten Ho 229* the *Gothaer Go 229*.

Walter Horten, in his Luftwaffe uniform and seen chewing on a cigar. He is showing a very serious face as he turns around to face the photographer. Post war he bristled each time someone called their all-wing *Horten Ho 229* the *Gothaer Go 229*. It was not *Rudolf Göthert* who labeled the *Horten's* turbojet powered flying machine the *Gothaer Go 229*. Instead, the label has been erroneously applied by contemporary Internet "aviation historians" who persist in this practice.

Reimar Horten shown at his Argentine ranch in the mid-1980s. This gentle man did not get excited over people mislabeling his all-wing *Horten Ho 229* masterpiece the *Gothaer Go 229*. Photo by author.

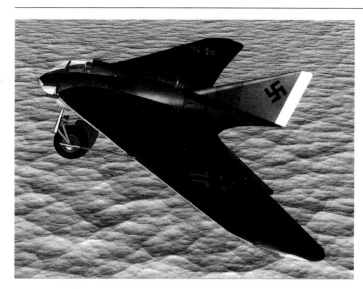

The proposed tailless Hort*en Ho 229* making its landing approach over snow-covered fields aided by its huge vertical stabilizer with an attached hinged rudder. Digital image by *Mario Merino*.

The overall planform of the proposed tailless *Horten Ho 229*. Digital image by *Mario Merino*.

The proposed Horten Ho 229 with its vertical stabilizer and hinged rudder as seen from its nose port side. Digital image by Mario Merino.

How the Horten Ho 229 might have looked with Walter Horten's proposed vertical stabilizer with an attached hinged rudder as viewed from its rear port side. Digital image by *Mario Merino*.

HORTEN HO 10A

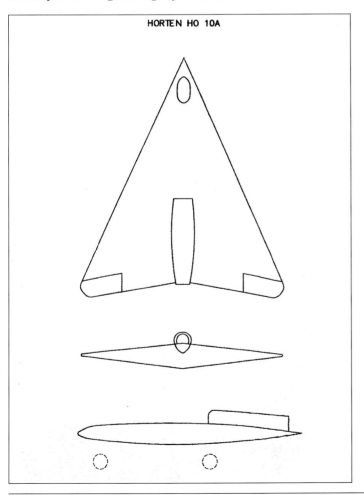

HORTEN HO 10B

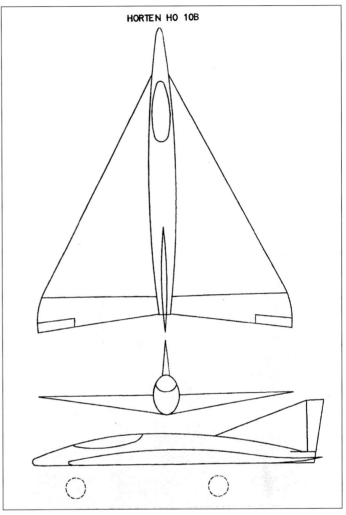

A poor quality pen and ink three-view drawing of the proposed delta-winged Horten Ho 10B. It would have featured a vertical stabilizer with a hinged rudder demanded by Walter Horten to counter criticisms by *Dr.-Ing. Rudolf Göthert* of *Gothaer Waggonfabrik* that *Horten* all-wing military flying machines, such as their *Horten Ho 229*, lacked directional stability and were therefore unsuitable for military use.

Left: A poor quality pen and ink three view drawing of the proposed delta winged *Horten Ho 10A* single turbojet engine and single seat *Luftwaffe* fighter. It would have lacked any vertical stabilizer with a hinged rudder. *Walter Horten* was opposed to their Horten Ho 10A because the proposal lacked a vertical surface, given the criticism being mounted by *Gothaer Waggonfabrik's Dr.-Ing. Rudolf Göthert*.

All-wing aircraft designer and constructor *Jack Northrop* as seen standing out front of one of his all-wing bomber aircraft.

Secretary of the Air Force *Stuart Symmington* **(1900-1986). This Texan was a large stockholder in** *Consolidated-Vultee* **aircraft manufacturing company of Fort Worth, Texas.**

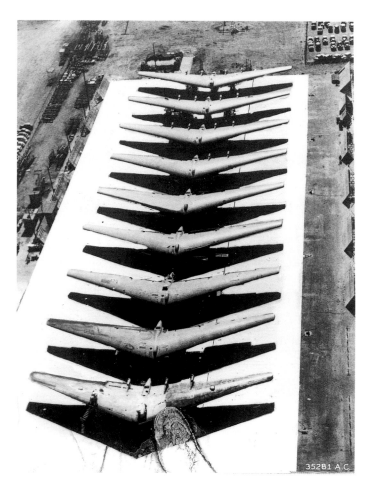

An outdoor Northrop Aircraft assembly line in southern California featuring nine all-wing *XYB-49* bombers undergoing their final fitting out.

The giant all-wing Northrop XYB-49 lifting off on its maiden flight in southern California.

A group of Luftwaffe officers. Left to right: *Walter Horten, Reimar Horten, Alexander Lippisch,* unknown, *General von Diesing,* and *Adolf Galland*. In the far right of the photo are *Wolfgang Späte* and *Rudolf Opitz*.

An early morning in southern California featuring the all-wing eight turbojet powered Northrop XYB-49 as viewed from its rear starboard side.

The Consolidated-Vultee (Convair) *B-36.* Seen from its starboard side. *Secretary of the Air Force Stuart Symmington, u*sed this flying machine to replace the all-wing *Northrop XYB-49*. The *B-36* was powered by six piston engines positioned along the wing's trailing edge and four turbojet engines, two mounted on a single pylon at each wing tip.

Pieces of a former Northrop XYB-49 cut up with acetylene torches as ordered by Secretary of the Air Force *Stuart Symmington.*

The Horten Ho 9 V1 sailplane version of the twin Jumo 004B powered Horten Ho 9 V2. This photo, featuring its starboard side, was taken in the Summer of 1944 by a *Horten Flugzeugbau* associate from the roof of its hangar at Göttingen.

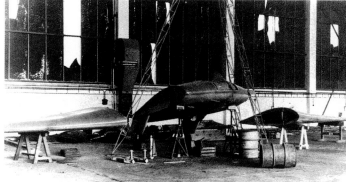

The disassembled Horten Ho 9 V1 as found by troops of the American 3rd Army at the Luftwaffe's Brandis/Leipzig Air Base in late March 1945. This all-wing sailplane is seen from its nose starboard side.

The highly deteriorated and abandoned center section of the *Horten Ho 9 V1* sailplane's center section as seen on the ground at *General George McDonald's Luftwaffe* collection center post war. It is believed that the entire *Horten Ho 9 V1* was considered scrap and recycled.

A ground level nose-on view of the like-new Horten Ho 229 V3's center section as delivered to Wright-Patterson Air Force Base, Dayton, Ohio, in late 1946.

A nose starboard side view of NASM's *Horten Ho 229 V3* center section. Photo by author.

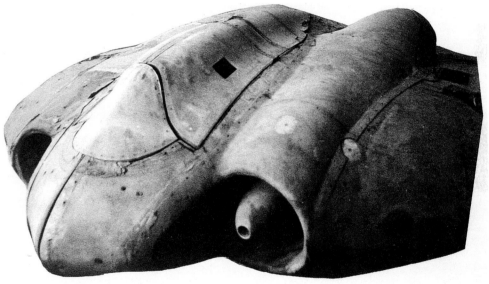

A nose port side view of NASM's Horten Ho 229 V3 center section. Photo by author.

A ground level direct nose-on view of NASM's dust covered *Horten Ho 229 V3*. Photo by author.

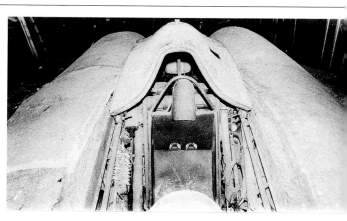

The open cockpit of NASM's Horten Ho 229 V3 looking aft. Photo by author.

A direct view into the open litter-filled cockpit of NASM's *Horten Ho 229 V3*. The pilot's seat is to the right, and the instrument panel is to the left in the photograph. Photo by author.

A direct tail-on view of NASM's Horten Ho *229 V3*. Photo by author.

A tail starboard side view of NASM's Horten Ho 229 V3's center section. Photo by author.

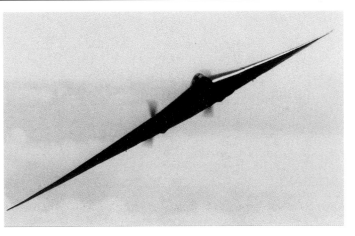

Twin Junkers Jumo 004B turbojet engines were buried in the *Horten Ho 229* series, and their thrust exited out over the center section's trailing edge. Featured is the yet-to-be restored *Horten Ho 229 V3* at the NASM, Washington, DC. Photo by author.

The aircraft design philosophy of *Walter* and *Reimar Horten* was to minimize performance robbing drag. This led to the all-wing sailplane planform in the early 1930s. As their designs evolved and they added piston motors, the H*ortens* buried everything within the wing. With their first turbojet powered all-wing machine only the air intake showed, as well as the exhaust tail pipe. Whether the two brothers recognized it or not, in the early 1940s, this was how fighter and bomber design would come to be...minimization of the flying machine's radar cross section. Shown is the twin piston motored all-wing *Horten Ho 7.*

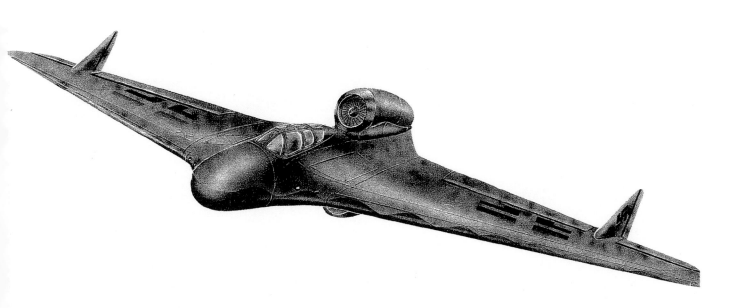

What legacy, if any, did Dr. Rudolf Göthert's proposed Gothaer Go P.60 series leave post war? At war's end Rudolf Göthert was thoroughly interrogated by *AAAFI*. Afterward, he was not offered any aircraft design work by the Allies and remained in Germany. His design work on the *Gothaer Go P.60* series, especially the tailless *Go P.60B* and *Go P.60C*, fell into obscurity and was later totally forgotten. One big reason is that aircraft designers have been continuously seeking to reduce the radar cross section of their fighter and bomber aircraft during the past fifty years since the end of World War II. The *Gothaer Go P.60B* and *Go P.60C* with their outside mounted over and under turbojet engines, as well as their wing tip vertical stabilizers with a hinged rudder presented too large a radar cross section. Shown is a pen and ink illustration of the *Gothaer Go P.60C*. Courtesy of "*Luftwaffe Secret Projects*: *Fighters 1939-1945.*"

A *Northrop B-2 "Stealth"* bomber seen in a banking turn.

A *Horten Ho 7* twin piston motored experimental all-wing fighter prototype in a banking turn in 1942.

A ground level nose-on view of a Northrop B-2 "Stealth" bomber.

A ground level nose-on view of a Horten Ho 7 twin piston motored experimental all-wing fighter prototype in 1942.

A ground level nose-on view of the Horten Ho 9 V1 experimental all-wing sailplane version of the turbojet powered Horten Ho 9 V2 of 1943.

A direct tail-on view of a *Northrop B-2 "Stealth"* bomber.

A direct tail-on view of the all-wing Horten Ho 9 V1 experimental sailplane version of the Horten Ho 9 V2 prototype all-wing prototype twin turbojet powered fighter.

A direct tail-on view of the all-wing Horten Ho 9 V2 all-wing twin turbojet powered fighter prototype of 1944.

A ground level starboard side view of a *Northrop B-2 "Stealth"* bomber.

A ground level starboard side view of the Horten Ho 9 V2 all-wing twin turbojet powered fighter prototype.

A ground level starboard side view of the *Horten Ho 9 V1* all-wing experimental sailplane version of the turbojet powered *Horten Ho 9 V2* of 1943.

An in-flight rear port side banking view of a Northrop B-2 "Stealth" bomber.

An in-flight rear port side view of a Horten Ho 7 all-wing twin piston motored fighter prototype of 1942.

A ground level port side rear view of the Horten Ho 9 V1 all-wing experimental sailplane version of the turbojet powered Horten Ho 9 V2 of 1943.

Port side profile (in-flight) of a *Northrop B-1 "Stealth"* bomber.

A starboard side profile of the Horten Ho 9 V2 all-wing turbojet powered fighter prototype of 1945.

A starboard side view of a *Northrop B-2 "Stealth"* bomber in low altitude level flight.

A starboard side view of the Horten Ho 5C experimental twin motored all-wing seen in low altitude level flight of 1942.

Index

Notes